Indian woman shelling acorns, to be ground into meal. (See page 70.)

Charles Francis Saunders

EDIBLE AND USEFUL WILD PLANTS

of the United States and Canada

Illustrated with Photographs
by the Author
and with Drawings
by Lucy Hamilton Aring

Dover Publications, Inc., New York

To

DOROTHY F. H.

Lover of Wild Things

This Volume

Is Affectionately Inscribed

Published in Canada by General Publishing Company, Ltd., 30 Lesmill Road, Don Mills, Toronto, Ontario.

Published in the United Kingdom by Constable and Company, Ltd., 10 Orange Street, London WC 2.

This Dover edition, first published in 1976, is an unabridged republication of the revised edition published by Robert M. McBride & Co. in New York in 1934 under the title *Useful Wild Plants of the United States and Canada.*

In this reprint edition, three illustrations by Stefen Bernath are substituted for inadequate pictures in the previous editions (pp. 118, 120, and 256).

International Standard Book Number: 0-486-23310-3
Library of Congress Catalog Card Number: 75-46193

Manufactured in the United States of America

Dover Publications, Inc.

180 Varick Street

New York, N.Y. 10014

INTRODUCTORY STATEMENT

ALL the familiar vegetables and fruits of our kitchen gardens, as well as the cereals of our fields, were once wild plants; or, to put it more accurately, they are the descendants, improved by cultivation and selection, of ancestors as untamed in their way as the primitive men and women who first learned the secret of their nutritiousness. Many of these—as, for example, the potato, Indian corn, certain sorts of beans and squashes, and the tomato—are of New World origin; and the purpose of this volume is to call attention to certain other useful plants, particularly those available as a source of human meat and drink, that are to-day growing wild in the woods, waters and open country of the United States. Though now largely neglected, many of these plants formed in past years an important element in the diet of the aborigines, who were vegetarians to a greater extent than is generally suspected, and whose patient investigation and ingenuity have opened the way to most that we know of the economic possibilities of our indigenous flora. White explorers, hunters and settlers have also, at times, made use of many of these plants to advantage, though with the settlement of the country a return to the more familiar fruits and products of civilization has naturally followed. Man's tendency to nurse a habit is nowhere more marked than in his stubborn indisposition to take up with new foods, if the first taste does not please, as frequently it does not; witness the slowness with which the tomato came into favor, and the Englishman's continued indifference to maize for human consumption.

Sometimes, however, the claims of necessity override taste, and there would seem to be a service in presenting in a succinct way the known facts about at least the more readily utilized of our wild plants. The data herein given, the writer owes in part to the published statements of travelers and investigators (to whom credit is given in the text), and in part to his own first hand observations, particularly in the West, where the Indian is not yet altogether out of his blanket, and where some practices still linger that antedate the white man's coming. The essential worth of the plants discussed having been proved by experience, it is hoped that to dwellers in rural districts, to campers and vacationists in the wild, as well as to nature students and naturalists generally, the work may be practically suggestive.

The reader is referred to the following standard

works for complete scientific descriptions of the plants discussed: Gray's *Manual of Botany of the Northern United States* (east of the Rockies); Britton and Brown's *Illustrated Flora of the United States and Canada* (the same territory as covered by Gray); Small's *Flora of the Southeastern United States;* Jepson's *Manual of the Flowering Plants of California;* Coulter's *New Manual of Botany of the Central Rocky Mountains;* Wootton and Standley's *Flora of New Mexico;* Piper and Beattie's *Flora of the Northwest Coast.* Abrams' *Illustrated Flora of the Pacific Coast,* projected in three volumes on the plan of Britton and Brown's work above mentioned, is also useful, but at this writing only one volume (covering the monocotyledons mainly and the willows) has appeared. For readers who do not find botanical keys intelligible it is hoped that the brief descriptions in the text will, with the aid of the careful line drawings, serve to identify most of the plants discussed.

TABLE OF CONTENTS

THE ILLUSTRATIONS IN HALF-TONE

THE ILLUSTRATIONS IN HALF-TONE

THE ILLUSTRATIONS IN LINE

THE ILLUSTRATIONS IN LINE

CHAPTER I

WILD PLANTS WITH EDIBLE TUBERS, BULBS OR ROOTS

Your greatest want is you want much of meat.
Why should you want? Behold the earth hath roots.

Timon of Athens.

THE plant life of the New World was always a subject of keen interest to the early explorers, whose narratives not only abound in quaint allusions to the new and curious products of Flora that came under their notice, but also record for many of our familiar plants uses that are a surprise to most modern readers. In that famous compilation of travelers' tales, published in England some three centuries ago under the title of "Purchas: His Pilgrimage," it is asserted of the tubers of a certain plant observed in New England that "boiled or sodden they are very good meate"; and elsewhere in Master Purchas's volumes there is note of the abundance of the same tubers, which were sometimes as many as "forty together on a string, some of them as big as hen's eggs."

GROUNDNUT
(Apios tuberosa)

This plant is readily identifiable as the Groundnut —*Apios tuberosa,* Moench., of the botanists—of frequent occurrence in marshy grounds and moist thickets throughout a large part of the United States and Canada from Ontario to Florida and westward to the Missouri River basin. It is a climbing perennial vine with milky juice and leaves composed of usually 5 to 7 leaflets. To the midsummer rambler it betrays its presence by the violet-like fragrance exhaled by bunchy racemes of odd, brownish-purple flowers of the type of the pea. Neither history nor tradition tells us what lucky Indian first chanced upon the pretty vine's prime secret, that store of roundish tubers borne upon underground stems, which made it so valuable to the red men that they eventually took to cultivating it about some of their villages. Do not let the name Groundnut cause you to confuse this plant with the one that yields the familiar peanut of city street stands, which is quite a different thing. The Groundnut is really no nut at all but a starchy tuber, which, when cooked, tastes somewhat like a white potato. Indeed, Dr. Asa Gray expressed the belief that had civilization started in the New World instead of the Old, this would have been the first esculent tuber to be developed and would have maintained its place in the same class with the potato.

Narratives of white travelers in our American wilderness bear abundant evidence to the Groundnut's part in saving them from serious hunger. Being a vegetable, it made a grateful complement to the enforced meat diet of pioneers and explorers; and Major Long, whose share in making known the Rocky Mountain region to the world is commemorated in the name of one of our country's loftiest peaks, tells in his journal of his soldiers' finding the little tubers in quantities of a peck or more hoarded up in the brumal retreats of the field mice against the lean days of winter. They may be cooked either by boiling or by roasting.

Though the Groundnut has so far failed of securing a footing in the gardens of civilization, there is another tuber-bearing plant growing wild in the United States that has a recognized status in the world's common stock of vegetables. This is a species of Sunflower (*Helianthus tuberosus*, L.), the so-called Jerusalem Artichoke. It is indigenous in moist, alluvial ground from middle and eastern Canada southward to Georgia and west to the Mississippi Valley, attaining a height at times of 10 feet or more. The French explorers in the St. Lawrence region in the early seventeenth century saw the tubers in use by the Indians and found them so palatable when cooked, suggesting artichokes, that they sent specimens back to France.

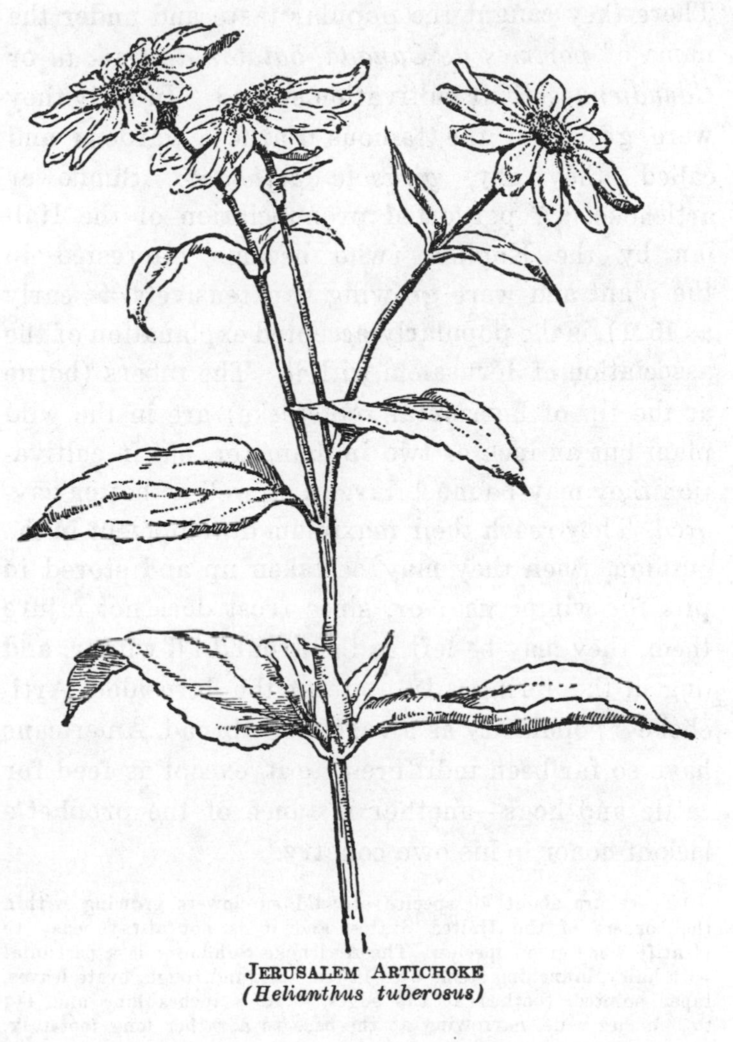

JERUSALEM ARTICHOKE
(Helianthus tuberosus)

There they caught the popular taste and under the name of *pommes de Canada, batatas de Canada* or *Canadiennes,* their cultivation spread. In Italy they were grown in the famous Farnese gardens and called, they say, *girasole articiocco,* Sunflower artichoke. A perverted pronunciation of the Italian by the English (who became interested in the plant and were growing it extensively as early as 1621), is the popularly accepted explanation of the association of Jerusalem with it. The tubers (borne at the tip of horizontal rootstocks) are in the wild plant but an inch or two in diameter, but in cultivation they may be much larger, as well as better flavored. They reach their maximum development in the autumn, when they may be taken up and stored in pits for winter use; or, since frost does not injure them, they may be left in the ground all winter, and dug in the spring. In spite of the Jerusalem Artichoke's popularity as a vegetable abroad, Americans have so far been indifferent to it, except as feed for cattle and hogs—another instance of the prophet's lack of honor in his own country.[1]

[1] There are about 40 species of wild sunflowers growing within the borders of the United States, and it is not always easy to identify some given species. The Artichoke Sunflower is a perennial with hairy, branching stems 6 to 12 feet tall, and rough, ovate leaves, taper pointed, toothed at the edges, 4 to 8 inches long and 1½ to 3 inches wide, narrowing at the base to a rather long footstalk.

EDIBLE TUBERS, BULBS OR ROOTS

Upon dry, elevated plains in and contiguous to the Missouri River basin ranging from Saskatchewan through Montana and the Dakotas southward to Texas, you may find, where the plough has not exterminated it, another famous wild food plant—the Indian Bread-root of the American pioneers, known to them also as Prairie Turnip and Prairie Potato, and to the French Canadians as *pomme de prairie* and *pomme blanche*. Botanically it is *Psoralea esculenta*, Pursh, and its smaller cousin *P. hypogaea*, Nutt. They are lowly, rough-hairy herbs, resinous-dotted, with long-stalked leaves divided into five fingers, and bearing dense spikes of small bluish flowers like pea blossoms in shape. The tuberous root, one to two inches in length, resembles a miniature sweet potato. Its nutritious properties were well known to Indians and such whites of other days as had any respect for the aboriginal dietary; and Indian women found a regular sale for it among the caravans of white traders, trappers and emigrants that traveled the far western plains in pre-railroad

Flowers yellow, both disk and rays, the latter numbering 12 to 20, and 1 to 1½ inches long. There is another species, *H. giganteus*, L., one form of which growing in moist ground in western Canada has thickened, tuber-like roots which are similarly edible. These are the "Indian potato" of the Assiniboine Indians. Mr. W. N. Clute, in "The American Botanist," February, 1918, noted that the prairie species, *Helianthus laetiflorus*, Pers., also bears tubers, which are little inferior to those of *H. tuberosus*.

times. The fresh tubers, dug in late summer, may be eaten raw with a dressing of oil, vinegar and

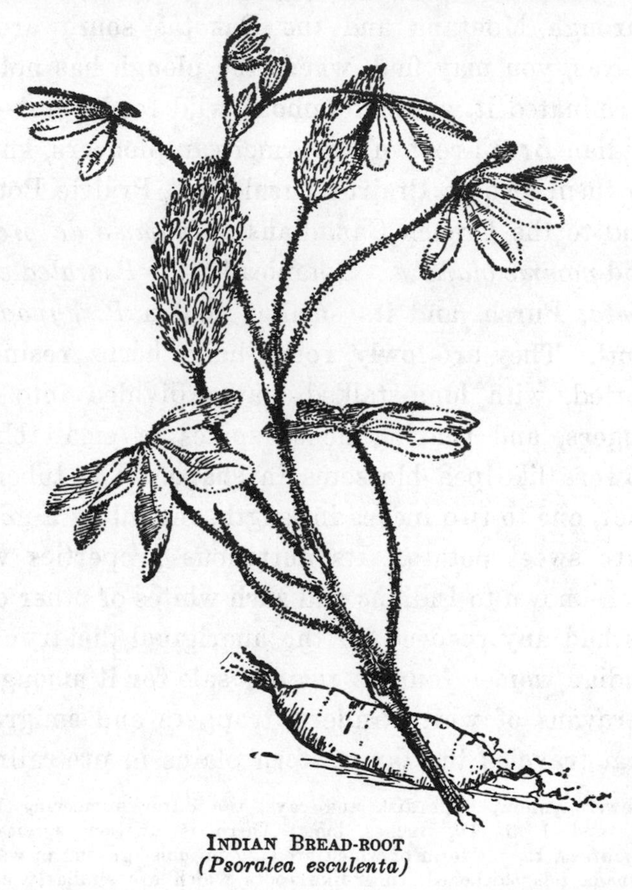

INDIAN BREAD-ROOT
(Psoralea esculenta)

salt, or they may be boiled or roasted. The Indians (who were habitual preservers of vegetable foods for winter use) were accustomed to save a portion of the Bread-root harvest, first slicing the tubers and then drying them in the sun or over a slow fire. The dried article was ground between stones and added to stews or soups, or mixed with water and baked in the form of cakes. The heart of the tuber is white and granular, and, according to an analysis quoted by Dr. Havard,[2] contains 70% starch, 9% nitrogenous matter and 5% sugar. Some attempts have been made to introduce it into cultivation as a rival of the potato, but the latter is so well entrenched in the popular regard that nothing has come of the effort. As a resource for those who are cut off from a potato supply, however, this free offering of Nature should be better known. John Colter, one of Lewis and Clarke's men, escaping from some Blackfeet who were intent upon killing him, lived for a week entirely upon these Bread-root tubers, which he gathered as he made his painful way, afoot, wounded, and absolutely naked, back to the settlements of the whites.

There are, by the way, two wild species of true potatoes indigenous to the mountains of New Mexico and Arizona—*Solanum tuberosum boreale,* Gray, and

[2] "Food Plants of the North American Indians," Bulletin Torrey Botanical Club, Vol. 22, No. 3.

S. Jamesii, Torr. The tubers are about the size of grapes, are quite edible when cooked and long ago attracted the attention of the Navajo and other Indians, who use them. And curiously in contrast to this the sweet potato of cultivation has a wild cousin in the United States (*Ipomoea pandurata*, Meyer) with a huge, tuberous root weighing sometimes 20 pounds, popularly called "man-of-the-earth." It is found in dry ground throughout the eastern United States, a trailing or slightly climbing vine with flowers like a morning glory. So obvious a root could hardly have escaped the Indian quest for vegetables, and as a matter of fact it was eaten to some extent after long roasting.

There is a plant family—the *Umbelliferae*—that has given to our gardens carrots, parsnips, celery and parsley. It includes also a number of wild members with food value, occurring principally in the Rocky Mountain region westward to the Pacific. Among these the genus *Peucedanum*, which by many botanists is now being called *Cogswellia*, is noteworthy because of the edible tuberous roots of several species. Of these the following may be noted, adopting Dr. Havard's enumeration in his paper above quoted: *P. Canbyi*, C. and R. (the chuklusa of the Spokane Indians); *P. eurycarpum*,

C. and R. (the skelaps of the Spokanes); *P. Geyeri*, Wats.; *P. ambiguum,* T. and G., *P. cous,* Wats. (the cow-as of the same Indians). The tubers may be consumed raw and in that state have a celery flavor. The most usual method of use among the Indians, however, was to remove the rind, dry the inside portion, and pulverise it. The flour would then be mixed with water, flattened into cakes and dried in the sun or baked. These cakes, according to Palmer,[3] were customarily about half an inch thick but a yard long by a foot wide, with a hole in the middle, by which they could be tied to the saddle of the traveler. The taste of such cakes is rather like stale biscuits. On

BISCUIT-ROOT
(Peucedanum Sp.)

[3] Edward Palmer, "Food Products of the North American Indians," Ann. Rept. U. S. Dept. Agriculture, 1870.

BISCUIT-ROOT
(Peucedanum ambiguum)

this account, the Peucadanums were commonly termed Biscuit-root by the white Americans. The Canadian French call them *racine blanche.* The genus is marked by leaves pinnate in some species, finely dissected in others, sometimes stemless and never tall, and with small white or yellow flowers disposed in umbels like those of the carrot or parsley. Novices, however, should be warned that the Umbelliferae include several poisonous species, and the investigator should be well assured of the identity of his plant before experimenting with it.

Then there is Yamp, of this same family, and cousin to the caraway. It is the botanists' *Carum Gairdneri,* H. and A.—a slender, smooth herb, sometimes four feet high, with scanty pinnate leaves 3- to 7-parted and white flowers like the carrot's, growing usually on dry hillsides in mountainous country from British Columbia to Southern California and eastward to the Rockies. The clustered, spindle-shaped roots are about half an inch thick, and raw have an agreeable, nutty taste, with a considerable sugar content. Not only Indians but white settlers also have proved the nutritive value of this root, eating it either raw or cooked. In meadows and along stream borders in Central California a nearly related species (*Carum Kelloggii,* Gray) frequently occurs and goes among the whites by the name of Wild Anise.[4] Its tubers are serviceable in the same way as Yamp. So also are those of *C. oreganum*, Wats., found from California to British Columbia. These are the *eh-paw* of the Klamath Indians, who regard them as a great delicacy.

A more famous root of the Pacific slope than Yamp is the Bitterroot (*Lewisia rediviva, Pursh*), the *racine amère* of the French explorers, and found from Arizona north to Montana (where it has given name to the Bitterroot Mountains and Bitterroot River) and west to the Pacific. It is a member of the Portulaca family, with showy, many-petaled white or pink blossoms sometimes two inches across and opening in the sunshine close to the ground, in form like a spoked wheel. Montana has adopted it as her State flower. It is one of the marvels in the history of alimentation that the unappetizing roots of this plant, intensely bitter when raw and smelling like tobacco when boiling, should have secured a stable place in any human bill of fare. Nevertheless, by the Indians of the far Northwest it has been extensively consumed from time immemorial, and explorers' journals contain many references to ab-

[4] Not to be confused with the mis-called Sweet Anise, which is really Fennel, *Foeniculum vulgare*. The latter is clothed with finely dissected leaves of a licorice flavor and has yellow flowers.

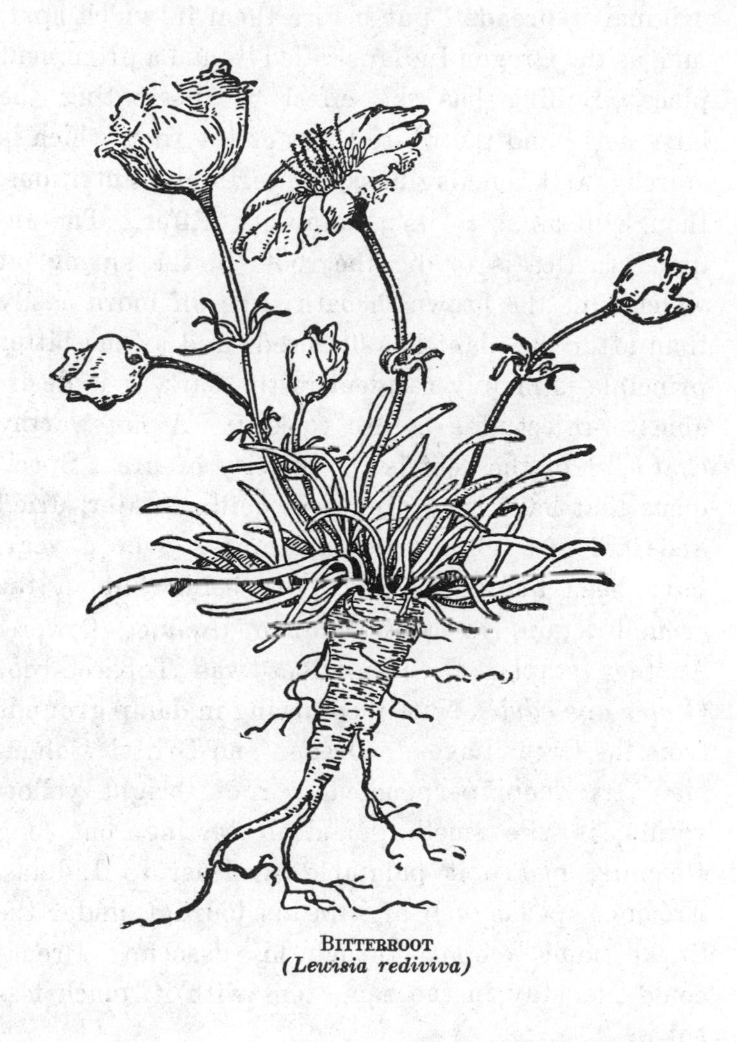

BITTERROOT
(Lewisia rediviva)

original "spreads" put before them in which spatlum, as the Oregon Indians called it, had a prominent place. Boiling has the effect of dissipating the bitterness; and the white heart of the root, which is starchy and mucilaginous, is certainly nutritious, though ideas as to its palatability differ. The Indian practice is to dig the roots in the spring, at which time the brownish bark slips off more easily than after the plant has flowered; and as the bitter principle is mainly resident in the bark, it is desirable to reject this before cooking. A noteworthy character of the root is its tenacity of life. Specimens that have been dipped in boiling water, dried and laid away in an herbarium for over a year, have been known to revive on being put in the ground again, to grow and to produce flowers. Another staple of some tribes was Tobacco-root (*Valeriana edulis*, Nutt.), occurring in damp grounds from the Great Lakes to Oregon and British Columbia. Its deep, perpendicular root, bright yellow within, is vile smelling and ill tasting, but long steaming makes it palatable, at least to Indians. Frémont speaks well of it in his journal, under the Snake name *kooyah*, though his associate Preuss could not stay in the same tent with it, much less eat it.

CHAPTER II

WILD PLANTS WITH EDIBLE TUBERS, BULBS OR ROOTS (*Continued*)

IT is a character of the Lily family that the plants are usually produced from subterranean bulbs or corms, and many such growing wild in the United States are of proved nutritiousness and palatability. Among these, for instance, are species of Allium, wild onion or leek, one of which particularly (*A. tricoccum*, Ait.) is recommended by those who have tried it for the sweetness and flavor of its young bulbs. It inhabits rich woodlands of the eastern Atlantic States north of South Carolina, its umbel of white flowers borne on naked stalks, appearing in June or July after its rather broad, odorous leaves have withered away. It is the Pacific Coast, however, that has a special fame for edible wild bulbs, many of which are known to the world at large only for the beauty of their flowers. There the Indians have, from before history began, been consuming such bulbs either raw or cooked. To some extent,

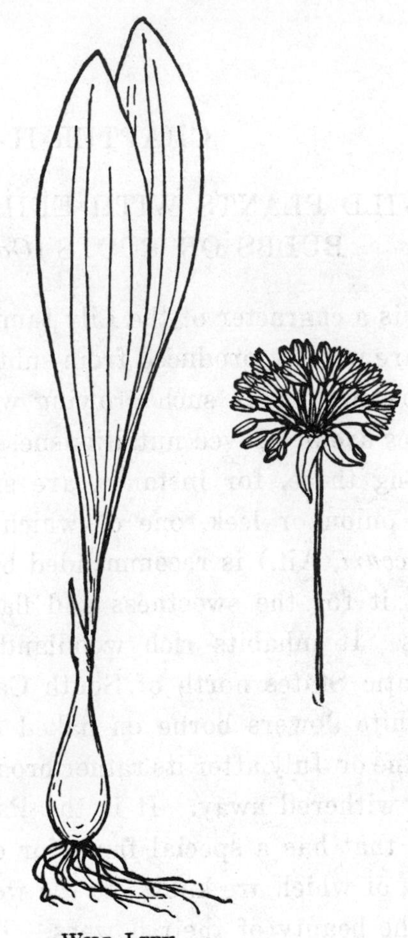

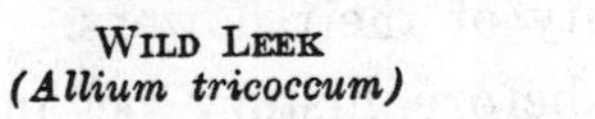

WILD LEEK
(Allium tricoccum)

Prickly Pear (*Opuntia Tuna*), one of the important food plants of the desert regions. (See page 108.)

also, they have been drawn upon for food by white travelers and settlers—the most palatable species being of the genera *Calochortus, Brodiaea* and *Camassia,* and commonly called "Indian potatoes." The genus Calochortus furnishes the flower gardens of both hemispheres with the charming Mariposa Tulips, and few who enjoy their beauty realize the gastronomic possibilities of the homely, farinaceous corms out of which the lovely blossoms spring. The species most widely known as a food source is *Calochortus Nuttallii,* T. and G., the Sego Lily, which has the distinction of being Utah's State flower. It may be recognized by its showy, tulip-shaped blossoms, whitish or lilac with a purple spot above the yellow heart of the

Sego Lily
(Calochortus Nuttallii)

flower, the leaves few and grass-like. It is indigenous to an extensive territory ranging from Dakota to Mexico and westward to the Pacific Coast. It was, I believe, a common article of diet among the first Mormons in Utah, under the name "Wild Sago," through a misunderstanding, perhaps, of the word "Sego," which is the Ute Indian term for this plant. A California species (*C. venustus*, Benth.) with white or lilac flowers variously tinged or blotched with red, yellow or brown, is also highly esteemed for its sweet corms. The cooking may be done by the simple process known to campers of roasting in hot ashes, or by steaming in pits, a method that will be described later on.

Brodiaea is a genus comprising numerous species, of which the so-called California Hyacinth, Grass-nut or Wild Onion (*B. capitata*, Benth.), common throughout the State, is perhaps the best known. Its clustered, pale blue flowers bunched at the tip of a slender stem are a familiar sight in grassy places in spring. The bulbs are about the size of marbles and noticeably mucilaginous. Eaten raw they seem rather flat at first, but the taste grows on one very quickly. They are also very good if boiled slowly for a half hour or so. The Harvest Brodiaea (*B.*

grandiflora, Smith), with clusters of blue, funnel-shaped flowers like little blue lilies, is another familiar species common in fields and grassy glades from Central California northward to Washington. Its bulbs are best cooked, as by slow roasting in hot ashes, which develops the sweetness.

Wild Onion
(Brodiaea capitata)

But the liliaceous bulb that has entered to the most important extent into the menus both of aborigines and white pioneers is the Camas or Quamash—"the queen root of this clime," as Father De Smet puts it in his "Oregon Missions." It is a handsome plant when in flower, which is in early summer. The 6-parted, usually blue blossoms, an inch or more across, occur in ample racemes at the top of stalks a foot or two high; the leaves all radical and grass-like. The bulb somewhat resembles a small onion, but is almost tasteless in the raw state. The range of the plant is from Idaho and Utah westward to central California, Oregon and Washington; and when undisturbed it grows so abundantly in open meadows and swampy lands as to convert them at a distance into the appearance of blue lakes of water. John K. Townsend, a Philadelphian who published an interesting narrative of a journey to the Rocky Mountains in 1839, has left us a pleasant, old-fashioned picture of a Camas feast in central Idaho. "In the afternoon," he writes, "we arrived at Kamas Prairie, so called from a vast abundance of this succulent root which it produces. The plain is a beautiful level one of about a mile over, hemmed in by low, rocky hills, and in spring the pretty blue flowers of the Kamas are said to give it a peculiar and very pleasing appearance. . . . We encamped here near a small branch of the Mallade River; and soon after all hands took their kettles and scattered over the prairie to dig a mess of Kamas. We were of course eminently successful, and were furnished with an excellent and wholesome meal. When boiled, this little root is palatable and somewhat resembles the taste of the common potato. The Indian method of preparing it, however, is the best."

This method, which embodies really the principle of our present day fireless cooker and has been employed by the aborigines from time immemorial for cooking numberless things, is briefly this: A hole of perhaps three feet in diameter and a foot or so in depth is dug in the ground and lined, bottom and sides, with flat stones. A fire of brushwood is then maintained in the hole until the stones are thoroughly heated through, when the embers are removed and fresh grass or green leaves (or, failing these, dampened dried grass) are spread upon the hot rocks and ashes. Upon this the bulbs are laid, covered with another layer of verdure or wet hay; and the whole is then topped with a mound of earth. In this air-tight oven the bulbs are left to steam for a day and a night, or even longer. The pit is then opened and the Camas will be found to be soft, dark brown in color, and sweet—almost chestnutty—in taste. The cooked mass, if pressed into cakes and then dried in the sun, may be preserved for future use.

There are several species of Camas, but the one best known is the botanist's *Camassia esculenta*,

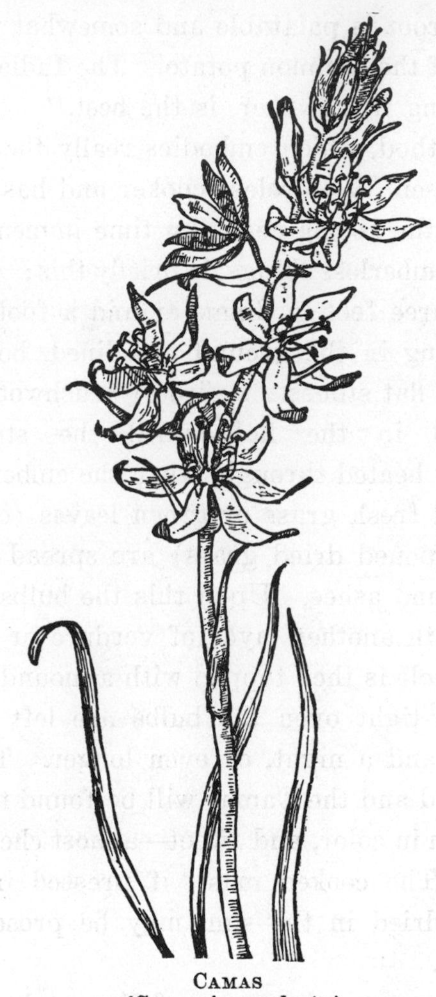

CAMAS
(Camassia esculenta)

Lindl., the plant of the preceding paragraphs. A closely allied species is *Camassia Leichtlinii* (Baker) Cov., common in northern California and Oregon. White settlers, in the days before their orchards and gardens were established, found in Camas a welcome addition to their meager and monotonous bill of fare, and Camas pie was a not uncommon dish in many an old time Oregon or California household.

Related to the Lily tribe is the Sedge family, of which two or three species are utilizable for human food. One of these is a bulrush of wide occurrence in the United States (*Scirpus lacustris,* L.), the Far Western form of which is commonly known as Tule. Its tuberous roots are starchy and may be ground, after drying, into a white, nutritious flour. They may also be chewed to advantage by travelers in arid regions as a preventive of thirst. Of more worth, however, are two species of Cyperus—*C. rotundus,* L., and *C. esculentus,* L. The former, commonly known as Nut-grass, is a denizen of fields in the Southern Atlantic States; the latter, popularly called Chufa, is abundant in moist fields on both our seaboards. (It is the *taboos* of the Mono Indians of California.) Like all of their genus, they are distinguished by triangular stems, naked except for a few grass-like leaves at the base, and bear-

CHUFA
(Cyperus esculentus)

ing at the summit of the stem an umbel of inconspicuous, purplish-green florets. The dietetic interest in them centers in the rootstocks, which bear small tubers of a pleasant, nutty flavor, and both white men and Indians have approved them, as well as the white men's pigs. The Chufa's hard tubers, especially, are sweet and tasty, and in some parts of the South have been considered worthy of cultivation, though by reason of rapid increase and difficulty to eradicate, the plant has a tendency to become a bad weed. We get the name Chufa from Spain, where the tubers are used in emulsion as a refreshment in the same class with "almonds in the milk, pasties, strawberries, azaroles, sugar icing and sherbets," according to some lines of a Spanish poem I ran across the other day.[1]

Of quite restricted occurrence in the United States, but worthy of mention because of its importance, is a member of a peculiar natural order of plants called Cycads. They resemble the palms in some respects and in others the ferns, their leaves, for instance, having a fashion of unrolling from base to apex in the manner of fern croziers. Many species inhabit tropical America, and two reach the southern

[1] "Almendrucos y pasteles,
Chufas, fresas y acerolas,
Garapiñas y sorbetes."

tip of our country, being indigenous to the Florida peninsula. One, known to botanists as *Zamia pumila,* L., occurs in dense, damp woods of central

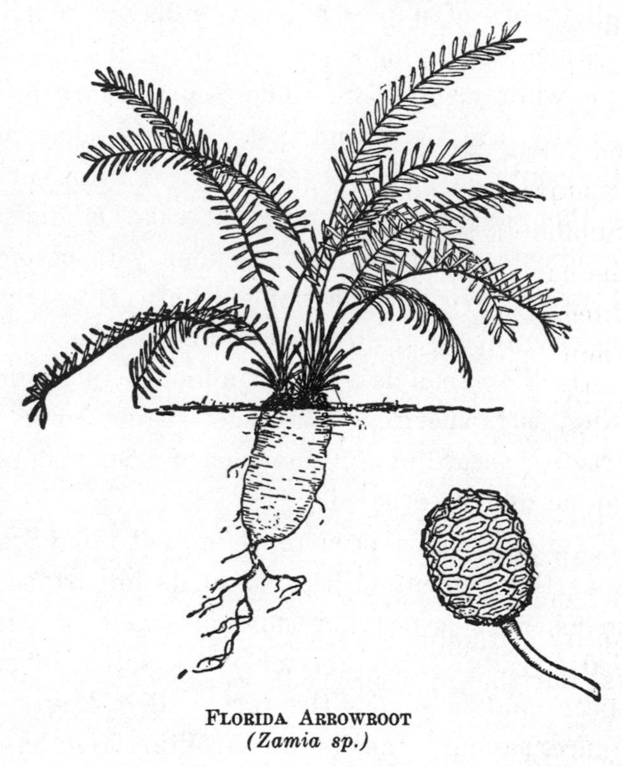

FLORIDA ARROWROOT
(Zamia sp.)

Florida: the other, *Z. Floridana,* DC., is a wilding of the open, dry, pine region of the east coast of southern Florida. They are popularly called Coontie or Coontah, the Indian name. The stiff, fern-like foliage arises in a clump from the crown (at the ground level) of a thick, subterranean stem which is exceedingly rich in starch. A nutritious flour made from the stem- and root-content of Zamia has had some vogue in the shops under the name of Florida Arrowroot. It has long been a staple article of diet with the Seminole Indians, and the plant has even found its way into the literature of juvenile adventure, as readers of boy romances may recall.

Similar in name to Coontie—indeed, probably the same name applied to a different food—is Conte or Contee, mentioned by William Bartram [2] as served to him by the Seminoles, and prepared from the starchy, tuberous roots of the China-brier (*Smilax Pseudo-China*, L.). This dish was made by chopping up the root, pounding the pieces thoroughly in a mortar, then mixing with water and straining through a sort of basket filter. The sediment was dried and appeared as a fine, reddish meal. A small quantity of this mixed with warm water and honey, says Bartram, "when cool, becomes a beautiful, delicious jelly, very nourishing and wholesome. They also mix it with fine corn flour, which, being fried in fresh bear's grease, makes very good hot

[2] "Travels through North and South Carolina, Georgia, East and West Florida, etc.," 1773, Chap. VII.

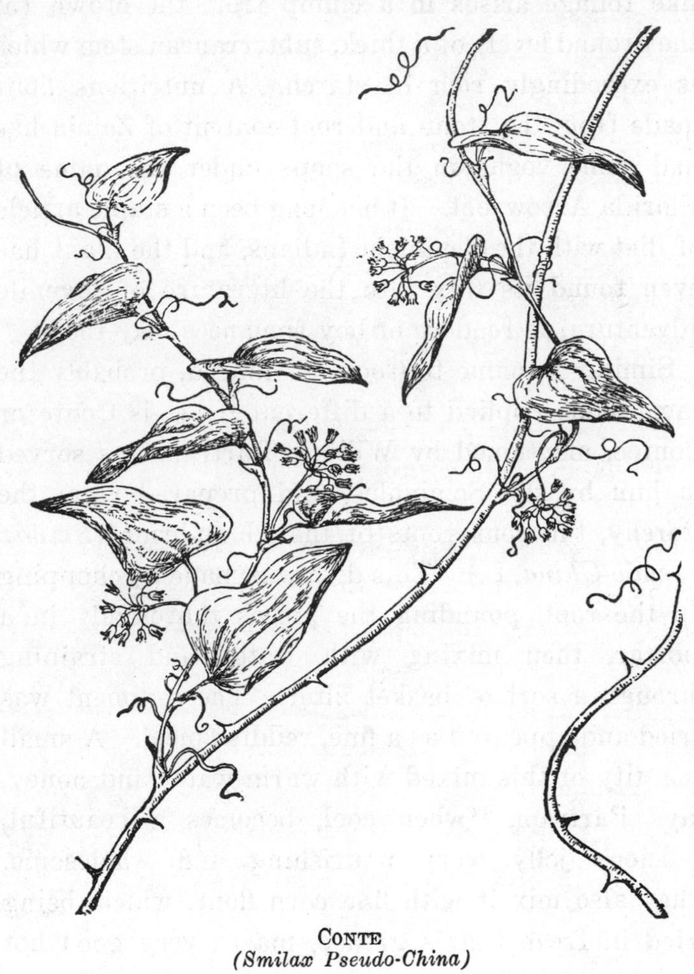

CONTE
(Smilax Pseudo-China)

cakes or fritters." So, you see, the wilderness as well as the town had its gastronomic delicacies, and dallied with dyspepsia. The China-brier, sometimes called Bull-brier, is a perennial woody vine of dry thickets from Maryland to the Gulf of Mexico, adorned in autumn with showy umbels of black berries not known to be edible. The whites have used the knotty, tuberous roots as the basis of a home-made rootbeer in association with molasses and parched corn.

Our waters, too, yield some native roots of economic worth. Among these aquatic wildings perhaps the commonest is the Arrowhead (*Sagittaria variabilis,* Eng.), so called from the shape of its leaves. It is found in swamps, ditches, ponds and shallow waters very generally throughout North America from the Atlantic to the Pacific and from Canada to Mexico, flowering in summer with 3-petaled white blossoms arranged in verticels of three. All Indians, whether of the Atlantic Slope, the Middle West or the Pacific Coast, have set great store by the plant because of its starchy, white tubers, somewhat resembling small potatoes, developed in autumn at the ends of the rootstocks. It is nearly related to a cultivated vegetable of the Chinese—*Sagittaria Sinensis,* a native of Asia.

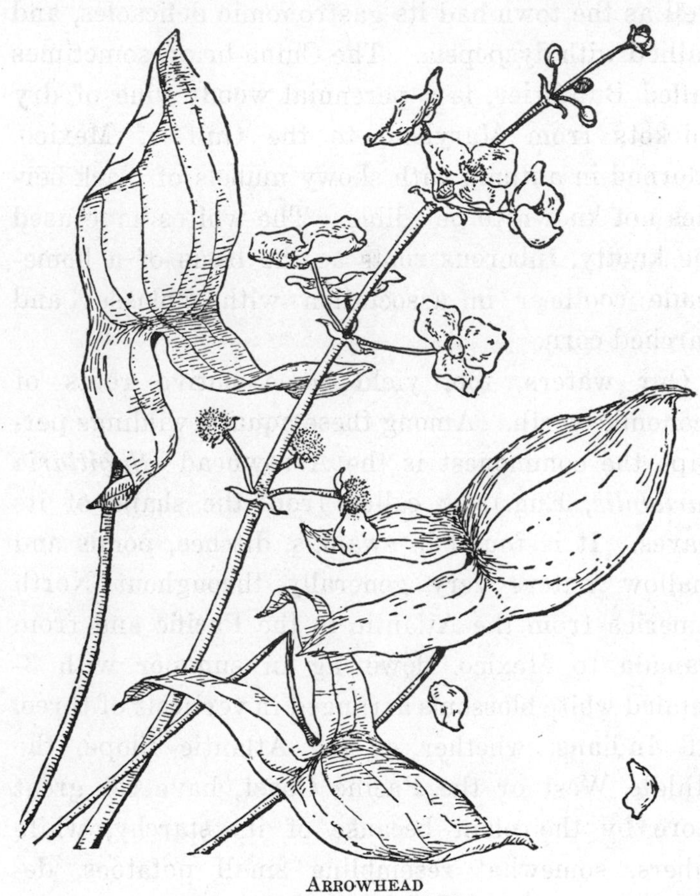

ARROWHEAD
(Sagittaria variabilis)

Lewis and Clark, in their narrative, speak of an island in the Columbia River, which they call Wappatoo Island, because of the numerous ponds in its interior abounding in the Arrowhead plant, which in the Indian language is termed Wappatoo. Those doughty explorers have given a picturesque description of the aboriginal Arrowhead business in the Columbia River country of Oregon as it was a century ago. "The bulb," to quote from their Narrative, "is a great article of food and almost the staple of commerce on the Columbia. . . . It is collected by the women, who employ for the purpose canoes . . . sufficient to contain a single person and several bushels of roots, yet so very light a woman can carry them with ease. She takes one of these canoes into a pond where the water is as high as the breast, and by means of her toes separates from the root the bulb which on being freed from the mud rises immediately to the surface of the water and is thrown into the canoe." Roasted or boiled, the tubers become soft, palatable and digestible, and to travelers in the wild make a fairly good substitute for bread. Tule potatoes they are sometimes called.

Also as bread upon the waters is that majestic aquatic, native to quiet streams and ponds of the interior United States from the Great Lakes to the

Gulf, the American Lotus or Water Chinquapin (*Nelumbo lutea,* Pers.). It is easily recognized by its huge, round leaves (sometimes two feet across and a favorite sunning place, by the way, for water snakes) lifted high above the water on foot-

WATER CHINQUAPIN
(Nelumbo lutea)

stalks attached to the center of the concave leaf, and its showy, pale yellow, papery flowers of numerous petals curving upward to be succeeded by curious, flat-topped, pitted seed-vessels. It is an American cousin of the famous lotus of India and oriental romance. To the American Indian, however, it seems never to have appealed as a flower of contemplation, but quite prosaically as an addition—and an important one—to his dinner table. In this rôle he found it trebly useful: first, because of the young leaves and footstalks which may be turned to account in the same way as spinach; secondly, because of the ripened seeds which, roasted or boiled, are palatable and nutritious with a taste that has given rise to the popular name Water Chinquapin; and thirdly, because of the large tubers, weighing sometimes half a pound each, which, when baked, are sweet and mealy with a flavor somewhat like a sweet potato. This is the plant whose flower is rather exuberantly referred to by Longfellow in "Evangeline":

> "Resplendent in beauty, the lotus
> Lifted her golden crown above the heads of the boatmen."

Though the customary habitat of this Nelumbo is the Mississippi basin, some isolated stations for it are known near the north Atlantic coast, notably in the Connecticut and Delaware Valleys, suggesting the view that it may have been introduced into such localities and cultivated by the Indian inhabitants. However the fact may be, its value as a food source is such as would have warranted such introduction.

The aroids—a plant family abundant in the tropics and of which several species, as the taro of the Pacific, possess nutritious, starchy, tuberous roots of importance as human foods—are represented in the United States by two or three plants of proved value. One of these is the Golden Club (*Orontium aquaticum*, L.), whose flower spikes of a rich, bright yellow, lifted above velvety, green, strap-like leaves from which water rolls as from a duck's back, are a familiar sight in the spring in ponds and marshes along the Atlantic coast. The bulbous rootstock, when cooked, is possessed of considerable nutriment, but owing to its deep seat in the muck is difficult of extraction. The ripened seeds, which resemble peas, are more easily gathered, and both whites and Indians have included them in their diet. According to Peter Kalm, an observant and inquisitive Swede whose book of travels in the North American Colonies in 1748 is still an interesting narrative to any who enjoy a look into the vanished past, the dried seeds, not the fresh, should be used, and they must be boiled and re-boiled repeatedly before they are fit to eat; yet his Swedish acquaintances thought it worth their while to do so.

Of even greater interest is another aroid, the Arrow Arum or Virginia Tuckaho (*Peltandra Vir-*

ginica, [L] Kunth, and perhaps the nearly related species *P. alba,* Raf., of the Southern States, a plant with large, arrow-shaped leaves and inconspicuous flowers enveloped in a green spathe. *Peltandra Virginica* is common in shallow waters of the Atlantic seaboard from Canada to Florida. I have never dug up the rootstock, about which I find the recorded descriptions differ. Havard, in his "Food Plants of the North American Indians," describes it, doubtless rightly, as short, deep-seated, sometimes six inches in diameter and weighing five or six pounds. As in the case of all aroids, the raw flesh of the rootstock is exceedingly acrid, indeed poisonous; but when dried and thoroughly cooked, it is found to have lost this objectionable principle, and in this state is a starchy food of proved nutrition. I think it is this plant that is meant in Purchas's Pilgrimage, where in the delicious English of the day record is made of the Virginians' "Tockawhough . . . of the greatness and taste of a potato, which passeth a fiery purgation before they may eate it, being poison whiles it is raw." The approved treatment appears to have been to steam it in the aboriginal heated pit, covered over with earth and left undisturbed for a day or two. Similarly the familiar Jack-in-the-Pulpit (*Arisaema triphyllum,* Torr.), whose small, turnip-shaped corm, bitten into raw, stings the tongue like red hot needles, becomes thoroughly tamed when dried and cooked, and its starchy con-

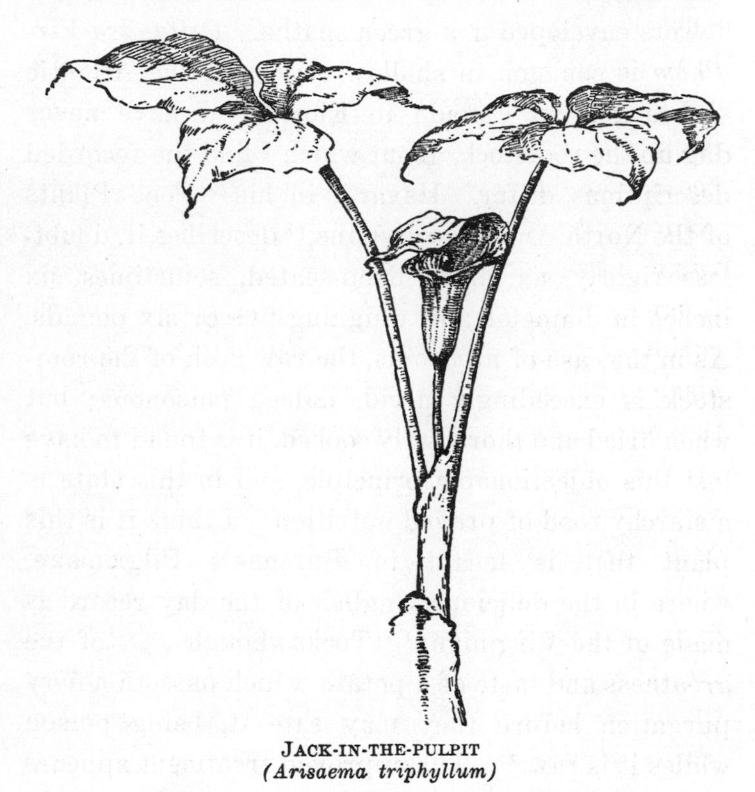

JACK-IN-THE-PULPIT
(Arisaema triphyllum)

tent was once a source of bread to the Seneca Indians.

The name Tuckaho has also been applied to a subterranean fungus (*Pachyma Cocos,* Fries), often found attached to old tree roots in the Southern States. It resembles roughly a cocoanut, though sometimes of more irregular shape. Inside the brown rind is a firm, white meat, which would be quite insipid, except for a trace of sweetness that is present. Its most common name is Indian Bread, because of the Indian use of it as a food. It is devoid of starch and seems of questionable nutritive value. Another subterranean parasite, though not a fungus, that is of genuine worth as an edible, is the curious Sand Food (*Ammobroma Sonorae,* Torr.), abundant in sandhills of southern Arizona and across the Mexican line in the dunes bordering on the Gulf of California, where it is called *camote de los médanos.* It consists underground of a slender, fleshy, leafless but scaly stem, two to three feet long, while above the sand during the flowering season in the spring is a small, funnel-like top on which the tiny, purple blossoms appear. After flowering, the overground part withers and disappears, and the plant presents no sign of its existence except to the experts who know where to dig. The subterranean stem is tender, juicy and sweet—a refreshing and luscious morsel, meat and drink in one. It may be eaten either raw or roasted, and is relished by redmen and white alike. Mr. Carl Lumholtz in his interesting book "New Trails in Mexico" tells of an Indian who lived almost entirely on Ammobroma, being able to find it out of season—a remarkable testimony to the nutritiousness of the plant and the abstemiousness of the Indian!

The creeping rootstocks of the common Cat-tail (*Typha latifolia,* L.) which covers great areas of our swamp lands throughout the United States, hold a nutritious secret, too, for they contain a core of almost solid starch. They were dug and dried in former times by Indians, who ground them into a meal. A recent analysis of such meal by one of the Government chemists showed it to contain about the same amount of protein as is in rice- and corn-flours, but less fat. It may make a useful mixture with the ordinary flours, and be substituted for cornstarch in puddings, as it seems entirely palatable.[3]

Thistles, proverbial feed for donkeys, are not usually listed among human foods, but both Lewis and Clark and Frémont found Indians eating the roots of at least two species. *Cirsium edule,* Nutt., common on the Pacific coast, was probably one, though this, I believe, owes its specific name to the edible young stems.

[3] The pollen of the flower spikes is also rich in nutriment and was made by some Indians into bread and mush.

CHAPTER III

WILD SEEDS OF FOOD VALUE, AND HOW THEY HAVE BEEN UTILIZED

The bounteous housewife, nature, on each bush
Lays her full mess before you.

Shakespeare.

THE Spanish conquest of Mexico and Peru brought to the knowledge of the white race a number of vegetable foods that are to-day on every American table—such as Indian corn, the potato, the pepper, and certain varieties of beans. Others are still unknown to the world at large. Among the latter that Cortés found in every-day use in Mexico was a square-stemmed, blue-flowered herb, which the chroniclers of that time called Chian or Chia. It seems to have ranked in popularity with staples like maize, frijoles, maguey, cacao and chili; and was grown with these in the fields and floating gardens of the Aztecs, for the sake of the small but numerous nutritious seeds of a pleasant, nutty flavor. Writers on the products of the New World in the first couple of centuries of the Spanish domination always speak of Chia with respect. Later, when upper California came in for settlement, the diarist of Portolá's expedition to the Bay of San Francisco specifies it as among the gifts offered by the Indians to their white visitors; and archæologists, grubbing in prehistoric graves in Southern California, have turned up deposits of the seed left as viaticum of departed souls, which attest the antiquity of its use within the limits of the United States. Even to-day, shopkeepers in the Spanish quarters of our own Southwestern cities as well as street venders in the towns of Mexico include Chia as part of their stock in trade.

One wonders what this all but forgotten food can be.

It is the name applied to at least five or six distinct species of plants, of somewhat different aspects, most of them belonging to the genus Salvia. The seeds are flattish and more or less shining, suggesting small flaxseed, of whose character they somewhat partake, being oily and mucilaginous. For human consumption they should be parched and ground, when they may advantageously be added to corn-meal, and this mixture made with water into a mush was a favorite item in the old Mexican dietary. Some of the present-day Indians of Southern California mix Chia meal with ground wheat, imparting to the latter a delicate, nut-like flavor, though the mucilaginous character of Chia disposes the mixture to gumminess. Pure Chia meal, mixed with water, cold or hot, swells to several times the original bulk, and is best eaten as a semi-fluid gruel. In this state it is soothing even to inflamed digestive organs. For desert people the meal constitutes an easily portable and highly nutritious ration eaten dry with the addition of a little sugar.

The species indigenous to the United States are *Salvia Columbariae,* Benth., and *S. carduacca,* Benth. Both are winter annuals native to the Pacific side of the continent. The former is the more common, found in dry ground throughout Southern California and adjacent parts of Nevada, Arizona and Mexico. The small, blue flowers, crowded in dense, prickly, globular heads, interrupted upon the stalk (which passes through the midst like a skewer), appear from March to June, and the seeds are ripe a month or so later. They are easily gathered by bending the stalks over a bowl or finely woven basket, and beating the heads with a paddle or fan, which shatters out the seeds. That is the Indian method; but when the plants grow plentifully, as

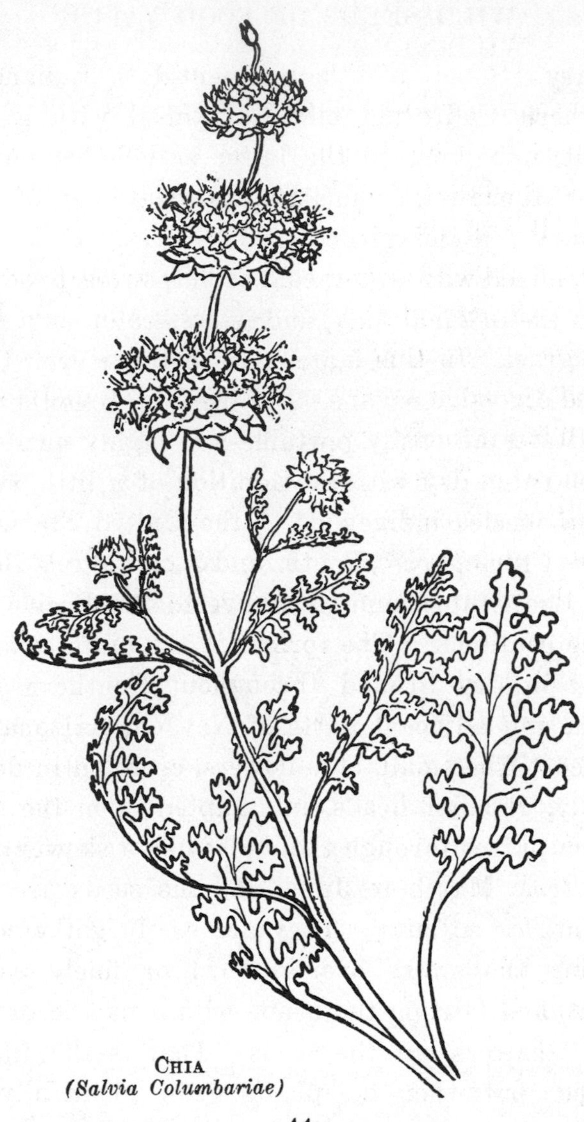

CHIA
(Salvia Columbariae)

they sometimes do as thick as grass in a field, or as they may be made to do by sowing the seed in cultivated ground, they can be cut, threshed and winnowed like flax or wheat.[1]

A wild food plant that has had a remarkable influence in geographic nomenclature is the Wild Rice (*Zizania aquatica*, L.). It is the *folle avoine* of the French voyageurs, and the *menómin* of the Northwest Indians, to one tribe of whom—the Menominees—it gave a name. Mr. Albert E. Jenks, whose exhaustive monograph, "The Wild Rice Gatherers of the Upper Lakes,"[2] is a mine of information about the plant, instances over 160 places (counties, townships, towns, railway stations, rivers, creeks, lakes and ponds) which have borne a name synonymous with this same Wild Rice. It is of the same family as the rice of commerce, and is a species of annual grass found growing by the acre, even the hundreds of acres, in ponds, swamps and still waterways, both fresh and brackish, in virtually every State of the Union east of the Rocky Mountains, and also in Japan and China. It is exceptionally abundant in the regions bordering on the Great

[1] An important use of Chia is as the basis of a soft drink. See the chapter on Beverage Plants.

[2] Printed in the 19th Ann. Report, Bur. Amer. Ethnology.

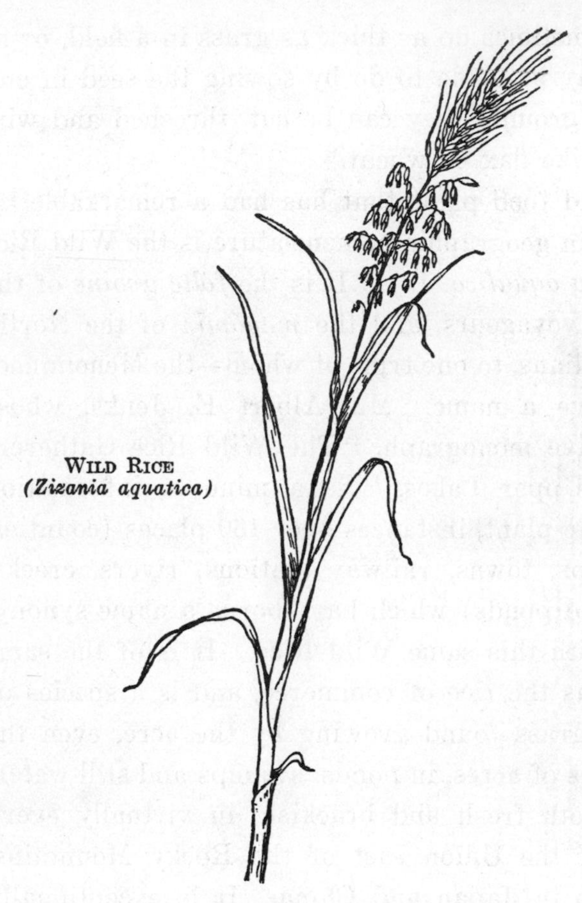

WILD RICE
(*Zizania aquatica*)

Lakes both in American and Canadian territory—a beautiful, stately grass, rising from two to twelve feet above the water and bearing in summer ample panicles of delicate, yellowish-green blossoms of two

An Indian of the Great Lakes Region threshing wild rice by means of a dasher-like stick.

(Courtesy of the New York Botanical Gardens.)

sexes. These are succeeded in September by the purplish spikes of ripened seeds occupying the tip of the panicle. The seeds are slender and cylindrical, one-half to three-fourths of an inch long, within a long-bearded husk and attached so loosely to the branchlet that bears them that they drop at a touch. They must needs be gathered, therefore, with great care or many may be lost. The Indians customarily harvest them just before they attain complete ripeness, visiting the rice swamps with canoes, which they push ahead of them, pulling the fruiting stalks over the hold of the canoe and beating the seeds into it with a stick.[3] The grain is then taken ashore where it is dried, either in the sun or by artificial heat upon racks under which a slow fire is kept burning. The husk must then be threshed off, which may be done by pounding with a heavy-ended stick in a bucket; and finally the chaff is got rid of by winnowing. The seeds are then ready for use or for storing away. Readers of old journals of the sojourners in the Northwestern wilderness will recall the important rôle played by such stores of Wild

[3] The best results are attained by first tying the standing stalks together at the head into small bunches. This is done a couple of weeks before maturity and serves to conserve the grain and lessen the depredations of the birds—particularly the bobolinks—which are famous rice eaters.

Rice (or Wild Oats, as the seed was as often but improperly called) in fighting hunger through the long, remorseless, northern winters.

The food value of Wild Rice is high. It is rich in carbohydrates (starch and sugar) and is also well stocked with flesh-producing proteids. Indeed, as a nutrient, it seems quite in the class of its cousin, the cultivated rice; and, like the latter, it swells with boiling, so that a little goes a long way. The Indians use it generally in mixture with stews. If cooked alone, two parts of water to one of rice is the usual proportion, and from a half to an entire hour is required for boiling it. White people who test Wild Rice usually pronounce it palatable, particularly in the form of a mush served with cream and sugar, and Mr. Jenks reports a wilderness soup made of Wild Rice and blueberries that sounds as if it ought to be good even in New York.

Two other water plants should be noted for their valuable edible seeds. One is the Water Chinquapin, mentioned in the previous chapter because of its useful roots, but which owes its popular name to the more obvious virtue of its palatable, nutlike seeds. These, boiled or baked, are considered by many the equal of chestnuts. The other is the Great Yellow Pond Lily of the northwestern Pacific Coast

(*Nymphaea polysepala* [Engelm] Greene), whose yellow flowers, sometimes as much as five inches in diameter, are a frequent and charming sight afloat on the bosom of shallow lakes and marshy ponds of the coast region from northern California to British Columbia. The globular seed vessels are full grown in summer, and it is the practice of the Indians to gather them in July and August, and, after drying the pods, to extract the seeds, which may then be kept indefinitely. These are commonly prepared for consumption by tossing them about in a frying pan over a fire until they swell and crack open somewhat as popcorn does, which they resemble in taste. They may be eaten thus out of hand, or ground into meal for making bread or mush.[4]

The common Sunflower of our gardens, whose monster heads appeal to esthetes because of a particular style of languid beauty they possess, and to birds and chickens because of their luscious, oleaginous seeds, is but a coddled form of one of our commonest wild plants—the Annual Sunflower (*Helianthus annuus*, L.). This species is indigenous throughout western North America, and sheets summer and autumnal plains for miles with the gen-

[4] Coville, "Notes on Plants Used by the Klamath Indians of Oregon." It was their characteristic food plant and called *wokas*.

erous gold of its cheery blossoms. The dark gray or blackish seeds of the wild plant are much smaller than those of the cultivated form, but are exceedingly numerous, with a white, oily, floury content that is rich in nutriment. They used to form an important part of the dietary of the Plains Indians, who sometimes cultivated the plants amid their corn. The ripe seeds were parched and ground into meal, and bread made of this meal has been spoken of with approbation by white travelers—even as the equal of corn bread. There can be no doubt of its value in situations where the flours of civilization are difficult to procure. As a source of oil sunflower seed is by no means insignificant, yielding, according to Havard, about twenty per cent. of an excellent table article. To most of us, indeed, the Wild Sunflower is a plant of unsuspected uses: its stalks possess a fibre of some worth and its flowers are good honey producers as well as a basis of a yellow dye said to be fast.[5]

In our Spanish Southwest the term *pinole* is in use

[5] *Helianthus annuus* is a coarse, much branched plant, three to six feet tall, the rough stem frequently mottled, the root (being annual) easily pulled up. The large flower heads are yellow-rayed with a dark center that is an inch or so across. Leaves petioled, ovate, six inches or more long, with toothed edges, rough to the touch. The seeds of the closely related species, *H. petiolaris*, Nutt., are similarly useful.

to mean meal made from the seeds of wild plants. Of these a great number have been utilized in past times for this purpose by the aborigines, and still are to some extent by old Indians whose taste for the pabulum of the long ago has not been lost. There is, it seems, a certain tang to the native vegetable foods of the wild comparable to the gaminess of wild flesh, that meets a need in untamed man not satisfied by the suaver products of civilization. The preparation of pinole is in a general way as follows: Provided with a large gathering basket of close weave and a paddle, usually of rough basket-work, the harvester beats the seeds—one sort at a time—into the basket, until a sufficient quantity is obtained. The chaff is then separated by sifting or by winnowing in a light breeze, and any prickles or hairiness natural to the seeds are singed off by dropping hot pebbles or live coals among them in a shallow basket and tossing all about at a lively rate. More prosaically, the same end may be attained with a frying pan kept agitated over a flame. This singeing process, moreover, serves to parch or partially cook the seeds, which are then ground in a mortar and the husks winnowed out. The residuum of meal, mixed with a little salt, may be eaten dry without further preparation. Indians in old times frequently made forced marches of a day on no other ration than a small sack of pinole, consumed in instalments as they traveled.[6] More often, however, it is moistened with water and eaten as mush or thinner as a gruel, or baked in the form of cakes. While the different sorts of seeds are collected and ground separately, it is not unusual to combine them for consumption, as taste may dictate.[7]

It would be tedious to enumerate all the plants which have been found of sufficient food value to grind into pinole, but the following may be mentioned as of especial interest and worth:

Of wide distribution in our Far West are two annual species of the homely Goosefoot or Pigweed. One is *Chenopodium Fremontii*, Wats., with more or less mealy leaves of triangular shape, a plant usually a foot or two high but sometimes attaining in overflowed lands a height of six feet or over; the other is *C. leptophyllum,* Nutt., with very narrow leaves that are scarcely mealy. The latter species occurs also in seashore sands of the Atlantic coast from Connecticut to New Jersey. The inconspicuous green

[6] For white consumption, the digestibility of this ration is improved by thorough and repeated grinding and parching after each operation.

[7] V. K. Chesnut: "Plants Used by the Indians of Mendocino Co., California." Printed as Contributions from the U. S. National Herbarium, Vol. VII, No. 3.

flowers of both species, clustered in panicled spikes, are succeeded in late summer and autumn by an abundance of small black seeds of farinaceous content. It stimulates our respect for these humble, weedy plants to know that the seeds of an allied species, *Chenopodium Quinoa,* have from the dawn of history been a valued food of the native Peruvians and Bolivians, and have been cultivated by those races. The Zuñi Indians of New Mexico, according to Stevenson, have a tradition that the seeds of *C. leptophyllum* were one of their principal foodstuffs in the infancy of the race before the gods sent them the corn plant. Afterwards, Chenopodium meal mixed with corn meal and salt, made into a stiff batter and moulded into balls or pats and steamed, became a favorite dish with epicurean Zuñis.[8] The seeds of a prostrate, mat-like Amaranth (*Amaranthus blitoides,* Wats.), a weedy plant with spikelets of greenish, chaffy flowers, native to the Rocky Mountain region and westward, also formed an important item in the ancient diet of the Zuñis, who believed that the original seeds of it had been brought up from the underworld at the time of the race's emergence into the light of day. In later years, the

[8] "Ethnobotany of the Zuñi Indians." 30th Ann. Report Bur. Amer. Ethnology.

meal made from these seeds has been used, like that from Chenopodium, in admixture with corn meal. Similarly useful to desert Indians are the seeds of species of Saltbush (*Atriplex canescens,* James, *A. lentiformis,* Wats., *A. Powellii,* Wats., *A. confertifolia,* Wats., etc.).

White Sage (*Audibertia polystachya,* Benth.), one of the most famous of Pacific Coast honey plants, produces slender, wandlike thyrses of pale blossoms whose seeds, though small and husky, are exceedingly numerous and rich in oil. They are still gathered by Southern California Indians, who bend the plants over a large basket and beat the seeds into it by striking with a seed-beater, as described before when treating of Chia. The seeds, mixed with wheat, are parched in a frying pan, and all is reduced to a fine meal by pounding in a mortar. This stirred in water with a sprinkling of salt is then ready to be eaten, or drunk, according as the mixture is thick or thin. It, too, is called *pinole.* The sage seeds have much the taste of Chia, the botanical relationship being close, but they are not mucilaginous.

Several species of wild grasses are utilizable for pinole. One of these is the Wild Oat (*Avena fatua,* L.), suspected of being the progenitor of the cultivated oat, and abundant in certain parts of the West,

Red Maple (*Acer rubrum*), the source of a dark blue dye in vogue among the Pennsylvania colonists. (See page 226.)

(*Courtesy of the New York Botanical Gardens.*)

particularly on the Pacific Coast where extensive areas are covered with it as with a crop. The seed resembles the cultivated grain, but is so hairy as to stick in one's throat and choke one. After thoroughly singeing off the hairs in a pan or basket tray, the grain may be reduced to flour, and used like ordinary oat-flour. Another pinole grass is *Elymus triticoides,* Buckl., locally known as "wild wheat" and "squaw grass." It is a tall, slim grass with usually glaucous stems, and grows densely in moist meadows and alkaline soil throughout the Pacific Coast and eastward to Colorado and Arizona. An allied species, more robust, with very dense flower-spikes of a foot long and larger seeds, serves a similar purpose. It is commonly called "rye grass" and is the *Elymus condensatus,* Presl., of the botanists. It, too, is abundant in damp, alkaline ground and along streams throughout the Far West, and Mr. Coville [9] has suggested that it may be worthy of experimentation as a cultivated grain for that region.

A Southwestern grass of wide distribution, particularly in the deserts, in sandy places (both moist and dry) and on arid hillsides, is the so-called Indian

[9] "Plants Used by the Klamath Indians," Washington, Gov't Printing Office, 1897.

Millet or Sand-grass (*Eriocoma cuspidata,* Nutt.). It is a perennial, growing in bunches a foot or two high, with peculiar panicles whose thread-like, twisting branchlets are tipped with husks containing small, blackish seeds, which have long been valued by desert Indians for flour making. This is one of the wild grains upon which the Zuñi Indians of New Mexico have been in the habit of relying in times of failure of their cultivated crops; and Dr. Edward Palmer tells of parties of Zuñis being seen as far as ten miles from their villages carrying enormous loads of these seeds for winter provision. Still another desert grass with edible seeds, but restricted in its distribution in our country to Southern California, is *Panicum Urvilleanum,* Kunth, which the desert Coahuillas call *song-wal.* It is a stout perennial, one to two feet high, the whole plant, including the seeds, more or less hairy, and is quite near of kin to the millet of the Old World, whose nutritious properties it shares.

Among the various gummy plants of the Pacific Coast known there as Tarweeds is one called Chile Tarweed (*Madia sativa,* Molina). It is a heavy-scented annual, one to three feet high, sticky and hairy, with rather narrow, entire leaves, and inconspicuous, pale yellow flowers of the daisy type, the rays barely a quarter of an inch long, expanding only at evening and early morning. This and some kindred species have been utilized by the California Indians for pinole. The Chile Tarweed has a special interest in the fact that in Chile, where it is also abundant, it has been cultivated from very early times. The seeds, when scalded, yield under compression a considerable percentage of a mild, agreeable oil, suitable for table purposes, soap-making, and notably for lubricating machinery, as it does not solidify short of 10° Fahr. Some eighty years ago, the plant was introduced into cultivation in Europe, where, I believe, it is still grown to some extent, and an oil-cake is made of the seeds for cattle.

To the traveler in the hill country of central and Southern California and western Arizona a familiar shrub is a species of wild plum with shining, evergreen, holly-like leaves (*Prunus ilicifolia*, Walp.), maturing in autumn an abundance of crimson or dark purple fruits in size and appearance like small damson plums. They are disappointing, however, in that they are almost entirely stone, though such thin covering of pulp as there is, is pleasant enough to the taste. It is an interesting fact in connection with the Indian's inventive genius that this fruit became long ago one of his important food sources; though it was not the pulp but the apparently hopeless pit that was turned to principal account. Gathering the plums in late summer, the Indians would

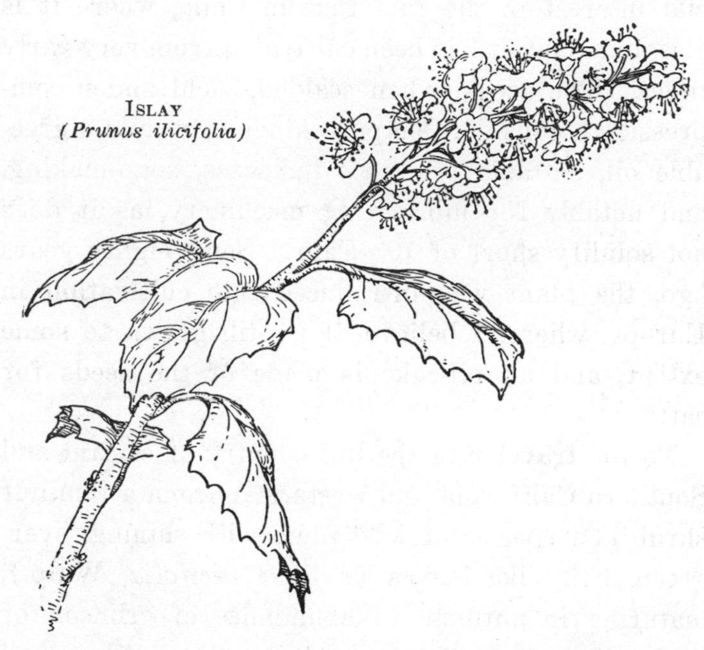

ISLAY
(Prunus ilicifolia)

spread them in the sun until thoroughly dry, when the stones would be cracked and the kernels extracted. These are bitter and astringent like acorns, and at first blush as unpromising as the uncracked pits themselves. When rid of that deleterious principle, however, the kernels are nutritious and digestible (by Indian organs, at least), and have always formed a cherished item in the native dietary, wherever the shrub grows. It is quite generally known by its Spanish-Indian name *islay.* Barrows, writing of this food,[10] states that the kernels are crushed in a mortar, leached in the sand basket (presumably like acorn-meal) and boiled as mush; but an intelligent old Indian of Mission Santa Inés, one Fernando Cárdenas, who is familiar with the customs practised by Southern California Indians, has informed me that the process as observed by him was to put the unground kernels into a bag and dip the sack in hot water again and again, until the meats became sweet. They were then ground, fashioned into balls and eaten so with great gusto. As I have personally never seen either process, I record both for the curious to test for themselves.

It would seem reasonable to expect edible seeds of many of the wild members of the useful Pea family, which is abundantly represented in all parts of the country. As a matter of fact, few seem to have been found worth while even by Indians of the most catholic taste. The Groundnut, *Apios tuberosa,* has been mentioned in a previous chapter as

[10] "The Ethnobotany of the Coahuilla Indians of Southern California." University of Chicago Press, 1900.

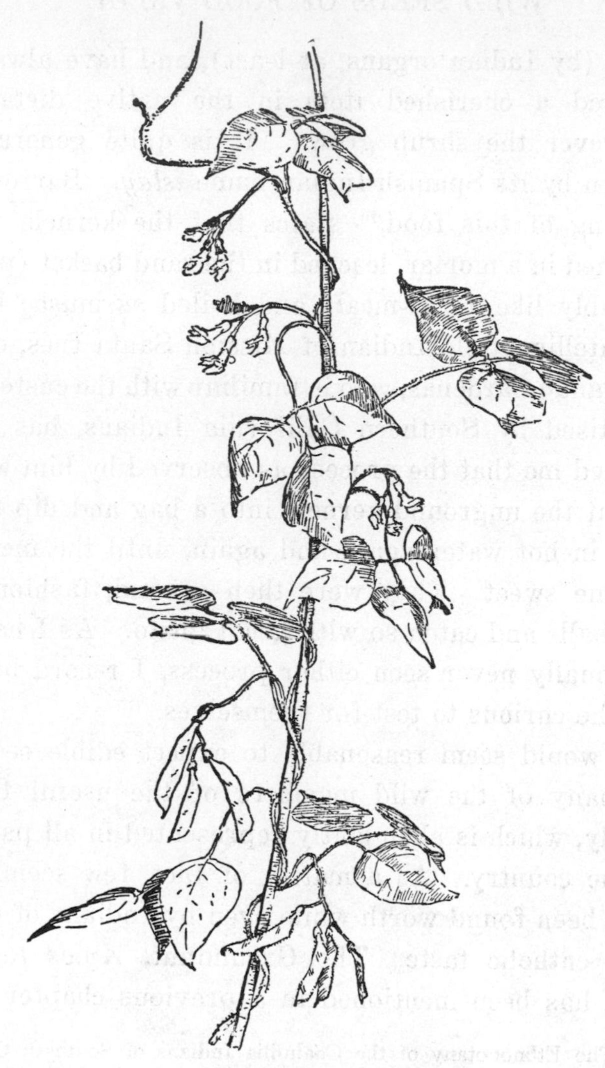

HOG PEANUT
(Amphicarpaea monoica)

having been utilized, both seeds and tubers; and something should be said of another leguminous plant popularly called Hog Peanut (*Amphicarpaea monoica,* Nutt.). It is a slender vine with trifoliate leaves, the stem clothed with brownish hairs, and is frequently met with in damp woodlands and thickets throughout the eastern half of the United States. In late summer it is graced with small bunches of pale purple or whitish pea-like blossoms, pendulous from the leaf-axils, while from near the root solitary, inconspicuous flowers on thread-like stems put out and bury themselves loosely in the ground, or creep shyly beneath a covering of fallen leaves. The showy upper blossoms are mostly abortive, though a few manage to develop short pods containing three or four small purple seeds apiece, edible when cooked. Of much greater worth are the subterranean seed-vessels which bear a single large pea in each. These peas are quite nutritious. They are mature in September and October, but retain their vitality throughout the winter, so that they may be dug even in the spring if one knows where to look for them.

The most valuable of all our wild legumes is doubtless the Mesquit-bean, the *algarroba* of the Mexicans. It is the product of a well-known tree

(*Prosopis juliflora*, DC., and its varieties) abundant throughout the arid region on both sides of the Mexican border. It is, indeed, the characteristic tree of the Southwestern deserts, giving to those gray wastes touches of living color very grateful to the eyes starving for the sight of a really vivid green. The pods, in shape and size resembling string beans, are produced abundantly in drooping clusters, which, ripening in late summer, become lemon yellow. The juicy pulp, in which the hard, bony seeds are embedded, is exceedingly sweet, containing, according to Havard, more than half its weight of assimilable nutritive properties, of which sugar is in the proportion of from twenty-five to thirty per cent. All stock thrives on the pods, and it is on this account rather than on any appeal to his own stomach that the white man's regard for them is grounded; but upon the Indian, who has ever a sweet tooth, they have a strong claim as human food. There is before me, as I write, a jar of coarse mesquit

MESQUIT
(Prosopis juliflora)

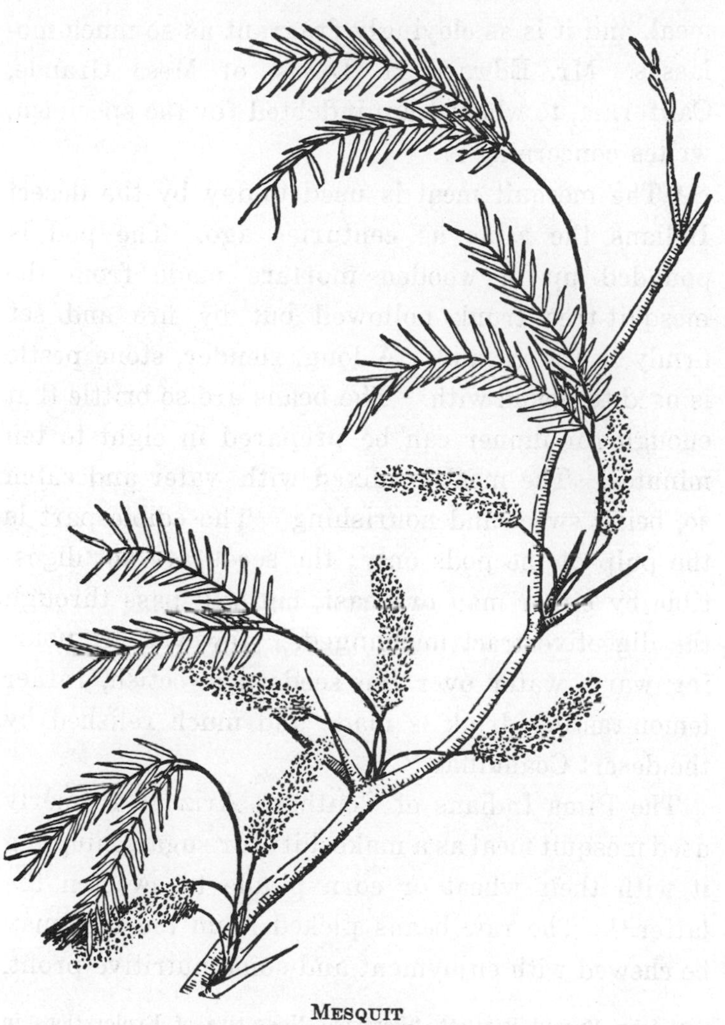

MESQUIT
(Prosopis juliflora)

meal, and it is as cloyingly fragrant as so much molasses. Mr. Edward H. Davis, of Mesa Grande, California, to whom I am indebted for the specimen, writes concerning it:

"The mesquit meal is used to-day by the desert Indians the same as centuries ago. The pod is pounded up in wooden mortars made from the mesquit-tree trunk hollowed out by fire and set firmly in the ground. A long, slender, stone pestle is used to pound with. The beans are so brittle that enough for dinner can be prepared in eight to ten minutes. The meal is mixed with water and eaten so, being sweet and nourishing. The edible part is the pulp of the pods only; the seeds are not digestible by either man or beast, but will pass through the digestive tract unchanged. However, by pouring warm water over the seeds a sweetish, rather lemon-tasting drink is made and much relished by the desert Coahuillas."

The Pima Indians of Southern Arizona formerly used mesquit meal as a makeshift for sugar, mingling it with their wheat or corn pinole to sweeten the latter.[11] The raw beans picked from the tree may be chewed with enjoyment and some nutritive profit,

[11] John Russell Bartlett, "Personal Narrative of Explorations in Texas, New Mexico, California, etc." Vol. II: 217.

as one travels. The quality of mingled acidity and sweetness which they possess before perfect maturity acts also as a thirst preventive, much as do the pods of the carob-tree of the Mediterranean basin. Indeed, the Spanish term *algarroba* current in Mexico (and in Hawaii where the tree is naturalized), is a case of transference, *algarrobo* being the word used in Spain for the carob-tree. A feature of the Mesquit-bean, by the way, to be reckoned with, is the fact that the pods are a favorite resort of a species of pea-weevil (*Bruchus*) for the deposit of their eggs. As a consequence Mesquit meal is particularly liable to infestation by these small beings to a degree that is somewhat of a shock to white sensibilities, though the Indians are indifferent to their presence; yet, I suppose, after all, it is no worse than skippers in over-ripe cheese. A gum that exudes from the bruised bark of the Mesquit resembles gum arabic with a similar, soothing quality. Moreover, boiled down it makes a black paint used by Arizona Indians to decorate pottery.

The Mexicans make a gruel, *atole de mezquite,* by boiling mesquit pods, mashing them in fresh water, and straining. A nutritious beverage is thus obtained. So altogether useful is the mesquit tree that it is not surprising to learn that it figures in the folklore of some regions where it grows. In Mexico a curious tradition is current to this effect: Long before the Spanish Conquest, the Apostle Thomas, in his heavenly home, became interested in the Aztecs, and descending to earth appeared to them in the guise of the Mexican hero-god Quetzacoatl and preached the gospel. The Aztecs heard the doctrine but coldly, and so San Tomás in most unchristian dudgeon departed, leaving the curse of sterility upon the plain of Anáhuac and turning all its cacao trees into mesquites, which remain mesquites to this day!

Closely related to the Mesquit-bean and of similar utility is the Screw-bean, called by the Mexicans *tornilla.* It is a curious, slender, spirally-twisted pod, borne in clusters, upon a small tree (*Prosopis pubescens*, Benth.) having much the same geographical range as the mesquit. The Screw-bean is even more sugary than the Mesquit-bean, and it may be made by boiling to yield a very fair sort of molasses. Water in which a small quantity of the meal is soaked makes a palatable and nutritious beverage. In making Screw-bean meal, the Indians grind the whole pods, seeds and all.

CHAPTER IV

THE ACORN AS HUMAN FOOD AND SOME OTHER WILD NUTS

Happy age to which the ancients gave the name of golden. . . . None found it needful, in order to obtain sustenance, to resort to other labor than to stretch out his hand and take it from the sturdy live-oak, which liberally invited him.

Don Quixote.

CERTAIN nuts growing wild in the United States, such as the chestnut, the hickories, the pecan, the beech-nut and the walnuts, have secured so firm a place in our civilized dietary that every one knows them, and they need not be discussed here. Perhaps, though, we have not exhausted all their culinary possibilities. For instance, William Bartram tells us that the Creek Indians in his day pounded the shellbark nuts, cast them into boiling water and then passed the mass through a very fine strainer. The thicker, oily part of the liquid thus preserved was rich like fresh cream, and was called by a name signifying "hickory milk." It formed an ingredient in much of their cookery, especially in hominy and corn cakes. Peter Kalm speaks of a similar practice observed by him with hickory nuts and black walnuts. A cooking oil is also said to have been obtained from acorns by some Eastern tribes, the nuts being pounded, boiled in water containing maple-wood ashes, and the oil skimmed off.

Of the nuts of our country unregarded by the white population from the standpoint of human food value, the noble genus of oaks supplies the most important. Every farmer realizes the worth of acorns for fattening hogs, but in America only the Indians, I believe, have taken seriously to utilizing them for human consumption; and it is significant that among the fattest of all Indians are those—the Californians—whose staple diet from prehistoric times has been acorn meal. There is, to be sure, a difference in acorns. All are not bitter. Several species of oak produce nuts whose sweetness and edibility in the raw state make it easy to believe the acorn's cousinship to the chestnut and beechnut. In this class are the different sorts of Chestnut Oaks, easily recognized by the resemblance of their leaves to the foliage of the chestnut tree; and of these perhaps the best, in respect of acorns, is *Quercus Michauxii,* Nutt.—commonly known as Basket Oak or Cow Oak. It is a large tree, indigenous to the Southern Atlantic

States in situations near streams and swamps, and ripening in September or October plump, sweet nuts an inch and a half long.

Oddly enough it is not the sweet acorns but the bitter that have played the really noteworthy part in aboriginal history. The Indians of the Pacific Coast did not become maize growers until after the white occupation of their country, preferring to accept from the hand of indulgent Nature such nutrients as came ready made, among which the abounding fruitage of extensive oak forests formed, and still forms, a conspicuous part. The acorns of all species of oaks indigenous to that coast are more or less stored with tannin, which imparts to the taste an unwholesome bitterness and astringency as disagreeable to red men as to white. Some inventive Indian—and doubtless it was a woman, the aboriginal harvester as well as cook—long ago hit upon a simple but effective way of extracting the deleterious principle; that is, washing the finely ground acorns in water. The process of preparing the acorn for human use, as still practiced in some parts of California, is as follows:

In autumn when the nuts are ripe but not yet fallen, they are gathered in baskets and barley sacks, brought home and laid in the sun to dry. Some are then stored away for future use in the house or in huge storage baskets set outdoors on platforms that are raised on legs above the reach of rodents, and form a picturesque feature of primitive rancherias. The acorns for immediate consumption are divested of the shells by cracking, and the kernels then reduced to the finest possible powder by grinding in the stone mortar, it having been found that digestibility depends upon thorough grinding.

The next step is to get rid of the bitterness, which persists through all the milling.

Every acorn-eating family maintains beside the nearest water a primitive leaching plant, varying more or less in the details of its make-up, but consisting primarily of a loose, concave nest of twigs, leaves or pine needles raised a foot or two above the ground and ensuring perfect drainage. Over this is stretched a piece of porous cloth—a clean burlap will do—sagging, basin-like, in the middle, upon which the meal is spread evenly about half an inch thick. Water, warm or cold, is then poured carefully over this and allowed to filter through, more being added from time to time until the bitterness is entirely leached away. The length of time required for this differs according to the variety of acorns used, some being less bitter than others.

A Western mountain Indian's storage baskets for preserving acorns and pine-nuts. They are elevated to forestall the depredations of rodents.

Two or three hours usually suffice. The result is a doughy mass, which is then transferred to a pot with water added, and boiled up for mush. It swells in cooking to about twice its original bulk, and when done is a pale chocolate color. In taste it is rather flat but with a suggestion of nuttiness that becomes distinctly agreeable even to some white palates. Judging from my own experience with it, I should pronounce it about as good as an average breakfast-food mush. Cream and sugar and a pinch of salt are considered needful concomitants by most white consumers. Formerly the Indians baked a sort of bread from acorn dough in their primitive fireless cooker—that is, in shallow pits first lined with thoroughly heated rocks. For this purpose the dough was usually, though not always, mixed with red clay in proportion of about five per cent., according to Mr. Chesnut, from whose valuable monograph, "Plants Used by the Indians of Mendocino Co., California," I have drawn for this statement,—the purpose of the clay being apparently to remove the last trace of tannin remaining in the dough. Upon a bed of green leaves placed at the bottom of the pit the dough was laid, covered with another layer of leaves, upon which a super-layer of heated stones was put, and all then covered with dirt, to remain over night. When removed after about twelve hours of slow cooking, the bread was coal black if the admixture of clay had been used or reddish brown otherwise, and of the consistency of soft cheese, hardening, however, with exposure. Such bread is oily and heavy, but noticeably sweet in taste. The latter characteristic is doubtless due to sugar developed by the prolonged, slow steaming.

Dr. C. Hart Merriam, in the "National Geographic Magazine" for August, 1918, tells of a simpler way of making acorn bread as observed by him. The hot acorn-mush is dipped, a small quantity at a time, from the general stock and plunged into cold water, which causes the lumps to contract and stiffen. The "loaves" so made are then placed on a rock to harden and dry out, after which they may be kept for weeks until consumed. The same authority speaks of the excellence of a bread made from a mixture of acorn-flour and corn-meal, in the proportion of one of the former to four of the latter.

While the acorns of any species may be utilized for human need, there is a distinct choice exercised by the Indians, the preference being based apparently on relative richness in oil and lowness in tannin. The best liked, according to my observation, are the Kellogg or California Black oak (*Quercus Californica*, [Torr.] Cooper), the Coast Live oak (*Q. agrifolia*, Nee), the Valparaiso or Canyon Live oak (*Q. chrysolepis*, Lieb), and the colossal Valley White oak (*Q. lobata*, Nee). An analysis of acorn meal made from the last named species is quoted by Chesnut as showing in percentage 5.7 protein, 18.6 fat, 65 carbohydrates (starch, sugar, etc.). Though the Californians have been rated among the lowest of our North American aborigines in native culture, their self-devised treatment of the acorn to make of it a wholesome food staple is entitled to the greatest respect. Stephen Powers, in his classic work on the Tribes of California, finds in one use of acorn mush an aboriginal discovery of the principle of the Prussian pea-sausage; and quotes the practice of a central California tribe, who, upon starting a journey, would pack in their burden baskets a quantity of the mush. When stopping for refreshment, it was only necessary to dilute a portion of this with water and dinner was ready. A woman, the traditional burden-bearer, could carry thirty pounds, enough to last two persons perhaps a fortnight. Naturally so important an element as the acorn in the tribal life became associated with religious ceremonial as well as incorporated in native poetry; and the approach of the autumnal gathering of the nuts was celebrated with dances and songs of thanksgiving and rejoicing. One of these songs, quoted by Powers, is Englished thus:

> "The acorns come down from heaven;
> I plant the short acorns in the valley;
> I plant the long acorns in the valley;
> I sprout, I, the black acorn sprout;
> I sprout."

Such dances (and they still have some vogue in the remoter parts of the State) were night affairs in the open, stamped out in the glow of blazing log fires to the accompaniment of minor melodies of fascinating appeal, the words of the songs repeated endlessly and emphasized with dramatic gestures, until the morning star appeared in the east. To this day the oak groves in those parts of California where any considerable Indian population still lingers are invested with traditional acorn rights, and recognized by general consent as the harvest grounds of particular communities, none poaching upon the preserves of another.

Traveling in mountainous regions of the West where coniferous forests prevail, one sometimes comes upon the remains of large camp-fires strewn roundabout with charred pine-cones and twig ends.

These are associated with another sort of nut[1] harvest, that of the Piñon or Pine-nut, the plump, oily seed of certain species of the Far Western pines. The most esteemed nut-pines are the Two-leaved Pine (*Pinus edulis*, Engelm.), a low, round-topped tree, generally known by its Spanish name *piñon* and common from Southern Colorado to Texas and westward to Arizona and Utah; the closely related One-leaved Pine (*P. monophylla*, Torr.), the piñon of the Great Basin region and desert slopes of the California Sierras; the Digger Pine (*P. Sabiniana*, Dougl.), a widely distributed species of the California foothills and lower mountain slopes; and the stately Sugar Pine (*P. Lambertiana*, Dougl.), whose huge cones are frequently a foot and a half long or more. The "nuts" of these species vary from one-half to three-quarters of an inch in length, with thin shells easy but rather tedious to crack. The meat is delicious in flavor even to white people, tender, sweet, and highly nutritious. They are, moreover, of easiest digestibility, so that even delicate stomachs are undisturbed by them. Under the name of *piñons* they are sold in towns throughout the Southwest as well as Mexico, where another

[1] The word "nut" is used in this chapter in its popular sense rather than with botanical accuracy.

species of nut-pine (*Pinus cembroides*, Zucc.) is indigenous. The Parry Pine (*P. quadrifolia*, Sudw.) is another good nut-pine, abundant in some parts of lower California, but only sparingly found on the United States side of the border. John Muir, in his picturesque way, characterizes the nut-pine forests as "the bountiful orchards of the red man."

Pine seeds are ripe in autumn, and the Indian method of gathering them is to cut or knock the unopened cones from the trees and then roast them in a camp fire. This serves to dry out the pitch and open the cones, from which the nuts are then easily extracted. The *piñon* harvest among the Southwestern Indians is a joyous time, and what they do not themselves consume is readily turned into money at the traders'. Dr. Edward Palmer, a veteran botanical collector whose notes are enlivened by many a human touch, describes a scene of this kind which he witnessed among the Cocopahs of Lower California. "It was an interesting sight to see these children of nature with their dirty, laughing faces, parching and eating the pine nuts . . . by the handful. . . . At last we had the privilege of seeing primitive Americans gathering their uncultivated crop from primeval groves." Though edible raw, the nuts are preferably toasted, which may be done very comfortably in a vessel kept in motion over a slow fire, as peanuts are heated. Not only is the flavor improved thereby, but the sweetness of the kernel is ensured for a longer time.

The value of the *piñon* was quickly recognized by the Spanish conquerors of New Mexico, and Fray Alonzo de Benavides in his famous Memorial to the King of Spain (1630) makes particular mention of the Piñon trees, marvelous to him "because of their nuts so large and tender to crack and the trees and cones so small and the quantity so interminable." It seems that at that early day there was trade in New Mexico piñons with the Mexican capital, a thousand miles away, where, Benavides tells us, they were worth at wholesale twenty-three to twenty-four pesos the fanega. They retail to-day in city shops of our Southwest at about twenty cents per pound.

In taking leave of the pines, a word should be said about the fruits of their cousins, the Junipers of familiar habit. Although reckoned as a conifer, the Juniper bears seed vessels that are not cones in the popular acceptance of that word, but berry-like, due to the growing together of the fleshy cone-scales, with a compact pulp around the seeds. The resinous quality of these "berries" in most species renders them repugnant to the human palate, but in a few cases this feature is much reduced and the "berries" are relished because of the sweet flavor of their mealy pulp. In this edible class are the fruits of the California Juniper (*Juniperus Californica*, Carr.), the Utah Juniper (*J. Utahensis*, Lem.), and the Check-barked or Alligator Juniper (*J. pachyphlaea*, Torr.). The first two are stunted trees or shrubs of arid regions of pure desert. The last is a tree attaining sometimes a height of fifty feet or more, abundant at rather high elevations in Arizona, New Mexico and Southwestern Texas, and remarkable for its thick, hard bark, deeply furrowed and checked in squares. The "berries" of all these species have been approved by Indian palates, and are eaten either raw or dried and ground into a meal and prepared as mush or cakes. Under necessity they might serve to keep body and soul together, those of the Alligator Juniper being considered the best. Cakes made from these are said on good authority to be palatable even to whites, and to have the merit of easy digestibility.

Little known to Americans but possessing a fascination all its own is the so-called Wild Hazel, Goat-nut or Sheep-nut, the fruit of a non-deciduous, grayish-green shrub, *Simmondsia Californica,* Nutt., locally abundant along the mountain borders of the

Jojoba
(Simmondsia Californica)

desert in Southern California and extending into Arizona and northern Mexico. It is a distant cousin to the beloved boxwood of old gardens, though none but a botanist would suspect the relationship. The plant is diœcious, so that not every individual is seed-bearing—only those possessing pistillate flowers. The capsules are mature in early autumn, and, gaping open, disgorge upon the ground the oily, chocolate-brown seeds, which are of about the size and appearance of hazelnut kernels. These, too, they somewhat resemble in taste, but are much easier of consumption because nature does the cracking for you. They are eaten with avidity by children, Indians, sheep and goats. Mexicans call them *jojobas,* and in Los Angeles I have seen them in the Spanish quarter in the shops of druggists, who find a steady sale for them for use in promoting the growth of deficient eyebrows! For this purpose, it seems, they are boiled, the oil extracted and this applied externally. The seed's reputation as a hair restorer, indeed, is rather extended in the Southwest. Mexicans in Lower California put it to still another use, which will be mentioned in the chapter on Beverage Plants.

According to M. Léon Dieguet in "Revue des Sciences Naturelles Appliquées" (October, 1895), "an analysis of the fire-dried seeds shows them to contain 48.30% of fatty matter. The oil solidifies at 5°, is suitable for food and of good quality, and possesses the immense advantage of not turning rancid." The shrub has been recommended for culture in the desert regions of the French Colonies of North Africa.

There is a beautiful little tree called the California Buckeye (*Aesculus Californica*, Nutt.) which whitens with its fine thyrses of bloom the hillsides of spring near streams in central and northern California. In summer and autumn it acquires another sort of conspicuousness due to the early dropping of its foliage, baring the limbs even in August. It then becomes a very skeleton of a tree upon which the fruits, hanging thick, look like so many dry, plump figs. The leathery rind of the latter encloses one or two thin-shelled nuts, shiny and reddish brown like those of the tree's cousins, the Buckeyes of the Middle West. To white folk these nuts, attractive as they appear, seem nevertheless devoid of food possibilities; indeed, in their raw state, they are known to be poisonous. That the Indian should have discovered how to turn them into fuel for the human machine seems, therefore, even more remarkable than the conversion of the acorn into an edible ration. Yet that is what the Indian did, by a method that consists essentially in roasting the nuts and then washing out the poison. One wonders how many prehistoric Californians died martyrs in the perfecting of the process. Mr. Chesnut, in his treatise already quoted on California Indian uses of plants, records in detail how the transformation into edibility is accomplished: The Buckeyes are placed in the conventional stone-lined baking pit which has been first made hot with a fire; they are then covered over with earth and allowed to steam for several hours, until the nuts have acquired the consistency of boiled potatoes. They may then be either sliced, placed in a basket and soaked in running water for from two to five days (depending upon the thinness of the slices), or mashed and rubbed up with water into a paste (the thin skin being incidentally separated by this process) and afterwards soaked from one to ten hours in a sand filter, the water as it drains away conveying with it the noxious principle. It was customary to eat the resultant mass cold and without salt. I have encountered no record of the similar use of the eastern Buckeye. The Californians' treatment of the Pacific Coast species is an interesting instance, I think, of what may be done with the most unpromising material.

CHAPTER V

SOME LITTLE REGARDED WILD FRUITS AND BERRIES

Greate store of forrest frute which hee
Had for his food late gathered from the tree.
The Faerie Queene.

NO one has to be told of the edibility of our wild strawberries, huckleberries, currants, cranberries, mulberries, raspberries, blackberries, elderberries, grapes and persimmons; nor of the pleasure which some palates find in the bitterish tang that goes with the familiar wild plums and cherries, although the only use to which most housewives consider these last fitted is the manufacture of jams and jellies. It is more to the purpose, therefore, in this chapter to touch upon some less known fruits of the hedge and heath—using the word fruit in its limited popular sense as based on succulency, rather than with botanical accuracy.

Throughout the basin of the upper Missouri and from Saskatchewan to New Mexico, the Buffalo-berry (*Shepherdia argentea,* Nutt.) is at home. In the journals of travelers in the upper plains two or three generations ago, no bush is more often men-

BUFFALO-BERRY
(Shepherdia argentea)

tioned than this. By the French *voyageurs* and *engagés* it was called *graisse de boeuf*, that is, "beef fat," which seems in harmony with the story I have read that the name Buffalo-berry is derived from the fact that it was a customary garnish to the monotonous buffalo steaks and tongue of those early days. The plant is a somewhat spiny shrub or small tree with silvery, scurfy leaves, and forms at times extensive and all but impenetrable thickets. The species is diœcious, and only the pistillate plant bears fruit; but that does it abundantly—tight clusters of small, scarlet berries, so sour as to find few takers until the frosts of October temper their acerbity. Then they are pleasant enough whether raw or cooked, though still with a touch of acid astringency that makes for sprightliness. Jelly made from them ranks especially high, and to this end they are gathered by white dwellers in the regions where they grow. In fact, the plant is not infrequently found transferred to gardens. The berries used to be one of the Indians' dietary staples, lending a lively, fruity flavor to the unending stews and mushes of the red men. There is a related plant, the Silverberry (*Elaeagnus argentea,* Pursh), native to much the same region and often cultivated in gardens for the sake of the fragrant, silvery, funnel-form flowers and attractive foliage. Its white, scurfy berries, while in a sense edible, are too dry and mealy for most people, and are left to the prairie chickens.

The Nightshade family, to which we owe the tomato, the potato and the egg-plant (as well as the tobacco and some very poisonous fruits), is represented in our wild flora by a number of plants bearing edible fruit. Of these the red berries of two shrubs of the deserts and semi-deserts of Arizona, New Mexico and Utah resemble tiny tomatoes and go among the Spanish-speaking population under the name of *tomatillo,* that is, "little tomato." They may be eaten raw, if perfectly ripe, or boiled and consumed either as a separate dish or used to enliven stews and soups. Dried, they look like currants and may be stored away for winter use. Botanically the plants are *Lycium pallidum,* Miers, and *L. Andersonii,* Gray. They are more or less spiny shrubs, with small, pale, narrowish leaves, bunched in the axils of the branchlets, and bearing funnel-form greenish or whitish flowers—those of *L. pallidum* nearly an inch long; of *L. Andersonii* much smaller. To the Navajo Indians, the berries of the former have a sacred significance and Doctor Matthews states that in his day they were used in sacrificial offerings to a Navajo demi-god. Similarly among the Zuñis the plant is sacred to one of their priestly fraternities, and treated with reverence as an intercessor with the gods of the harvest. When the berries appear, certain individual plants are sprinkled with sacred meal and this business-like prayer proffered: "My father, I give you prayer meal; I want many peaches."[1]

To the same family belongs the genus *Physalis,* some, perhaps most, species of which yield fruits that may be eaten. They are distinguished by a bladdery calyx which loosely envelops the small, tomato-like berry. These plants are known to Americans as Ground Cherries, and to the Spanish-speaking residents of our Southwest as *tomates del campo,* that is, "wild tomatoes." Of the score or so of species indigenous to the United States, *Physalis viscosa,* Pursh, is one of the best known—a hairy, sticky perennial, common in fields east of the Mississippi from Ontario to the Gulf. The nodding, greenish-yellow flowers have a purplish-brown center; and the yellow fruit is reported on excellent authority to be the best. A species producing red fruit (*P. longifolia,* Nutt.), found wild from Nebraska to Texas and westward to Arizona, has been thought worthy of cultivation by the Zuñi Indians, who used to grow it, and perhaps still do, in the women's quaint little gardens on the slope of the river Zuñi—

[1] Stevenson, "Ethnobotany of the Zuñi Indians." 30th Ann. Rept. Bur. Amer. Ethnology.

TOMATO DEL CAMPO
(Physalis longifolia)

gardens familiar to every observant visitor at this famous old pueblo. A favorite method of using the berries, according to Stevenson,[2] was to boil them and crush them in a mortar with raw onions, chili and coriander seeds. Among the whites, the Ground Cherries, when used at all, are made into preserves.

In the Rose sisterhood—a family that has given us a wealth of garden fruits—are a number of wildings of more or less food value. Next to the wild strawberries, raspberries and blackberries, none perhaps stands higher in popular favor than the Amelanchier, in popular parlance Service-berry, June-berry, Shad-bush or Sugar-pear.[3] It is found with specific variations in leaf and fruit on both our seaboards, as well as in the Middle West, a small tree or shrub with rather roundish, serrated leaves, and producing in late spring or early summer loose clusters of round or sometimes pea-shaped, crimson or dark-purple berries. These are juicy, with a pleasant taste not unlike huckleberries. To white settlers throughout the continent this berry has

[2] "Ethnobotany of the Zuñi Indians."

[3] *Service-berry,* a name transferred from an English species of *Pyrus,* whose fruit was known as *serb, serve* or *service; June-berry,* because the fruit generally ripens in June; *Shad-bush,* because blooming when the shad are running in Eastern rivers.

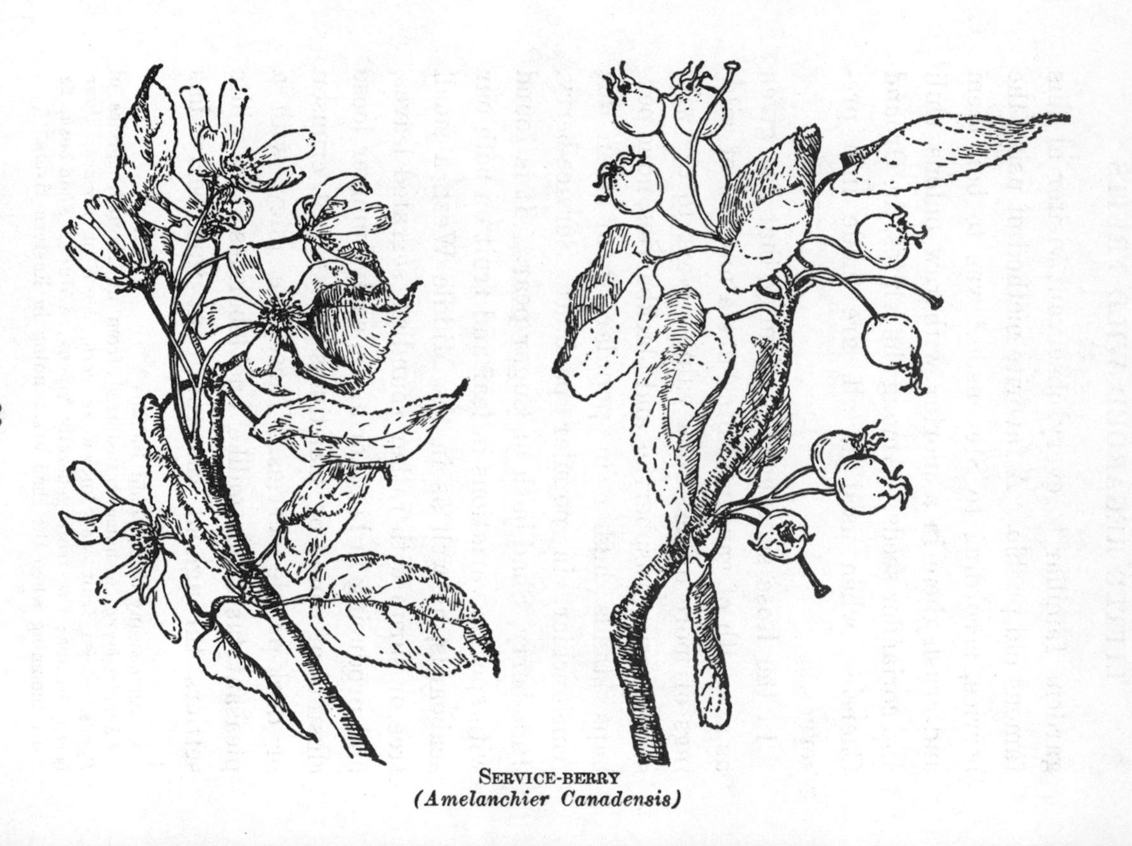

SERVICE-BERRY
(*Amelanchier Canadensis*)

A Southwestern desert hillside, which, in spite of its desolate look, bears plants yielding food, soap, textile fiber and drinking water. The man in the foreground is cutting mescal.

always been an abundant wild stand-by for fruit pies. Old time Indians used it not only fresh but dried for winter consumption. Lewis and Clark's journal mentions a berry that is undoubtedly this, which the Indians were observed preserving by pounding masses together into "loaves" of ten to fifteen pounds weight. These would keep sweet throughout the season and would be used as needed by breaking off pieces to be soaked in water and dropped into stews. Strong competitors with man for the berries are the birds and the bears.

Another western berry that has appealed strongly to Indian tastes but not, so far as I know, to ours, is the fruit of a species of Buckthorn (*Rhamnus crocea,* Nutt.). Doubtless there is nutrition in the berries, but they possess, according to Dr. Edward Palmer, the peculiar faculty of temporarily tingeing red the body of one who consumes them in quantity. He tells a gruesome story of accompanying as surgeon a troop of United States soldiers in pursuit of a band of twenty-two Apache Indians in Arizona, who were eventually surprised in their camp and killed outright. The bodies of all were discovered to be beautifully reticulated in red from the juice of the Rhamnus berries on which the Indians had been gorging, the color having been taken up by the blood and diffused through the smallest veins.

Our American Hawthorns (botanically, *Crataegus,* a genus which some modern botanists have split up into a hopeless multitude of confused species) bear clusters of tiny, alluring apples in various colors—yellow, purple, scarlet, dull red, some almost black. Many of these are admirable for jelly making. Among the best are the large haws of *Crataegus mollis* (T. & G.) Scheele, about an inch in diameter and of a bright scarlet color. The species is fairly common throughout the eastern United States and Central West. The Summer Haw (*Crataegus flava,* Ait.), a small tree of the Southern States, bears somewhat pear-shaped, yellowish fruits, one-half to three-fourths of an inch in diameter, which are also esteemed for jellies, as are the shining blackish berries of the Black Haw (*Crataegus Douglasii,* Lindl.), common in the Pacific Northwest, and sweet and juicy enough to be pleasant eating uncooked. In fact, when it comes to providing raw material for the jelly makers, almost any thicket in late summer will yield something, for even the hips of the Wild Rose have been turned advantageously to that use. The Spanish Californians gathered the ripe fruits (called by them *macuatas*) of *Rosa californica,* stewed them in a little water, added sugar, if they had it, and served with a dressing of the liquor they were cooked

AMERICAN HAWTHORN
(*Crataegus mollis*)

in. Or they would eat them raw from the bush, after frost had sweetened them.

On the Pacific Slope one of the cherished berries for jelly making is the Manzanita (*Arctostaphylos* of several species), a remarkable evergreen shrub, or sometimes a small tree, whose shiny, chocolate-colored trunk and twisting branches, as hard as bone, are familiar to every traveler in the California mountains. The popular name is Spanish for "little apple," and aptly describes the appearance of the fruit. This is borne very abundantly and is ripe in mid-summer. The mountain folk, describing the plant, will tell you there are two kinds, one with smooth berries and the other with sticky ones: but botanists are not so easily satisfied, and have described more than thirty species. The one most often used for jelly is *Arctostaphylos Manzanita,* Parry, common in mountainous regions throughout the length of California, and also, I believe, in parts of Arizona and Utah. The berries are smooth skinned, with an agreeable acid flavor, and nutritious, but dry, mealy and seedy. Chewed as one travels, they are a capital thirst preventive, but the pulp should be very sparingly swallowed, as it is quite hard to digest. Indians, in former days, however, set great store by them as an article of diet, and in specific Manzanita tracts, just as in the oak-groves, there were recognized tribal or family

MANZANITA
(Arctostaphylos Manzanita)

rights. The berries were consumed either dried and ground into pinole, or cooked as a mush, or in the fresh state. Death from intestinal stoppage is said to have sometimes resulted, however, from too free indulgence in the uncooked fruit.[4] A favorite aboriginal use, too, was in the manufacture of cider, which will be described in the chapter on Beverage Plants.

To white cooks the Manzanita is of negligible interest except, as already hinted, as a basis for a jelly, which is famously good. The following recipe I have from Mr. Edmund C. Jaeger of Riverside, California: Select berries, by preference of the smooth-skinned variety, which are more juicy than the others, picking them when full grown but still green, say about the first of June. Put them in a boiler with cold water to cover; and after bringing them to a boil, let them simmer until thoroughly cooked through: then pour into a cheese-cloth sack and press out the juice. This will have a cloudy look. Add sugar in the proportion of pound for pound, and boil till the liquid jells. The sugar clarifies the juice, and the jelly is a beautiful, clear, amber red. Should the berries be too ripe, there will be

[4] Chesnut. "Plants Used by the Indians of Mendocino Co., California."

failure to jell, but an excellent table syrup is the result, instead.

Wild currants, gooseberries, plums and cherries all play into the jelly maker's hands; and so do the acid, scarlet berries of the eastern Barberry (*Berberis Canadensis,* Pursh), found in mountain woods

OREGON GRAPE
(Berberis aquifolium)

from Virginia to Georgia, as well as of the European Barberry (*B. vulgaris,* L.) which has become a wild plant in some sections. On the Pacific slope another Barberry is the familiar Oregon Grape (*Berberis aquifolium,* Pursh), a shrub two to six feet high, with evergreen pinnate leaves of seven to nine leathery, holly-like leaflets, abundant in rich woods among rocks, especially in northern California and Oregon, of which latter State it is the floral emblem. Erect clusters of small but conspicuous yellow

OREGON GRAPE
(Berberis aquifolium)

flowers adorn the bushes in the spring, succeeded in autumn by blue berries of a pleasant flavor which are useful for jelly making and also as the basis of a refreshing drink. Cousin to the Barberry is the familiar May Apple, Wild Lemon or American Mandrake (*Podophyllum peltatum*, L.), a common herb, with umbrella-like leaves sheeting the ground in rich

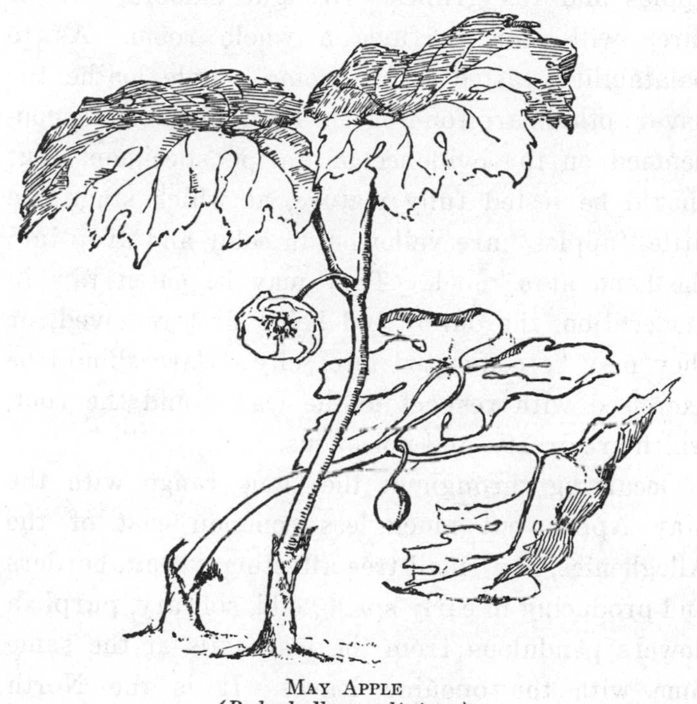

MAY APPLE
(Podophyllum peltatum)

woodlands and shady meadows throughout the region east of the Mississippi from Canada to the Gulf. The pear-shaped fruit, about the size of a butternut, has claims to edibility. When green it exhales a rank, rather repulsive odor, but when fully matured, all that is changed into an agreeable fragrance, hard to define—a sort of composite of cantaloupe, summer apples and fox grapes. Brought indoors, two or three will soon perfume a whole room. As to palatability, tastes differ: some people loathe the flavor; others are fond of it. It ought not to be condemned on the evidence of unripe specimens, but should be tested fully mature, at which stage the little "apples" are yellowish in color and drop into the hand at a touch. They may be eaten raw in moderation, the outer rind being first removed, or they may be converted into jelly. Care should be exercised with respect to the leaves and the root, which are drastic and poisonous.

Occurring throughout the same range with the May Apple, but much less common east of the Alleghenies, is a small tree affecting stream borders and producing in early spring odd, solitary, purplish flowers pendulous from the leaf axils at the same time with the opening leaves. It is the North American Papaw (*Asimina triloba,* Dunal). In September or October it bears sparse bunches of oblong, greenish, pulpy fruits each four or five inches in length and an inch or two in diameter, known as papaws, wild bananas, or, by old time French settlers, *asimines*—a Gallicized form of the Assiniboine Indian name of the fruits. They are unquestionably of some food value, though again tastes differ on the point of their palatability. "Edible for boys" is the classing they get from one good authority; but, on the other hand, the sweet, aromatic flavor is distinctly pleasant to some maturer palates. Perhaps, as I have heard it suggested, the divergence in views may be due in some degree to the fact of different natural varieties within the species. Our Papaw is a far-strayed member of the tropical family that includes the Anonas—the cherimoya, the sour-sop and the custard apples. Another plant tribe of the tropics that finds a small representation in the United States is the Passion Flower family, noted for its remarkable blossoms in which the devout have thought to see a perfect symbol of the Divine Passion. There is one species, commonly called Maypop (*Passiflora incarnata,* L.), so frequent along fence rows and in cultivated fields of the Southern States as to be in the class of a weed. The fruit is a yellow, egg-shaped berry, a couple of inches long, accounted edible, but more esteemed when made into jelly than when eaten raw. Nevertheless to some tastes the flavor is agreeable. I fancy it is to this plant that John Muir refers in his "Thousand Mile

Walk to the Gulf," quoting for it a local Georgia name, "Apricot vine," having a superb flower "and the most delicious fruit I have ever eaten."

The Heath family, which gives us the huckleberry, blueberry and cranberry (too well known to be treated here), as well as the manzanita already described, has two or three other members growing wild and bearing berries whose edibility is touched with a special grace of spiciness. One of these is the familiar Teaberry, Checkerberry or Wintergreen (*Gaultheria procumbens,* L.), an aromatic, creeping, evergreen vine usually of coniferous woods, from subarctic America southward through the eastern United States to Georgia. The crimson-coated berries, about the size of peas, are pleasant morsels and make a welcome feature in a small way in the autumnal displays of fruit venders in Eastern cities. A Pacific Coast species of Gaultheria with black-purple berries (*G. Shallon,* Pursh) has become commonly known by the name of Salal, a corrupted form of its Indian designation. It is a small shrub, one to three feet high, with sticky, hairy stems, frequent in the redwood forests of Northern California, and thence northward in shady woods as far as British Columbia. Lewis and Clark's journal contains several references to the Oregon Indians' fondness

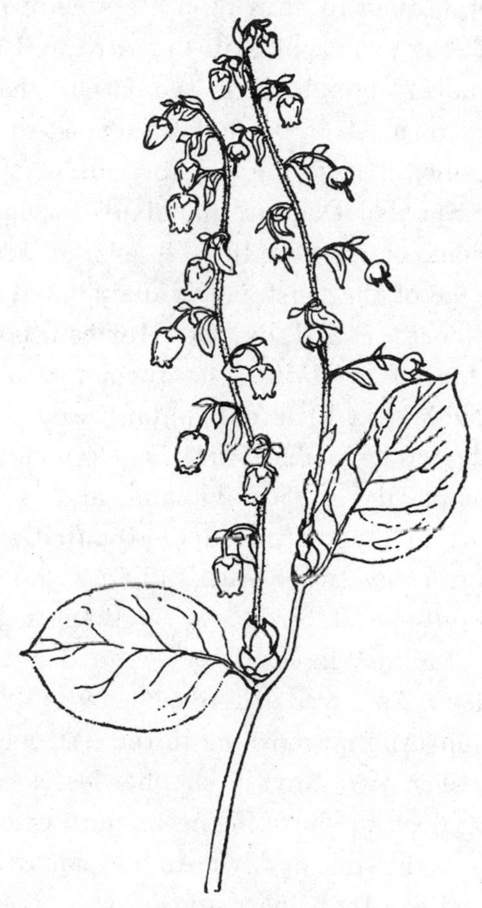

SALAL
(Gaultheria Shallon)

for the berries, which, under the names of Shallon and Shewel, seem to have been a staple of diet with them. Though thick of skin they are well flavored.

Paradoxical enough, it is the desert that grows some of our most important and most juicy wild fruits. Among these the plump pods of species of Yucca or Spanish Dagger, abundant throughout the arid regions of the Southwest, are of recognized worth. One of the most widely distributed is *Yucca baccata,* Torr., called by the Mexican population *Palmilla ancha* or *Dátil*—the former name meaning "broad-leaved little date-palm," and the latter, "the date fruit." The fruit is succulent, plump, and in shape like a short banana, and is borne in large, upright clusters, seedy but nutritious. The taste is agreeably sweet when fully developed, which is in the autumn if birds and bugs spare the pods so long. Indians have always regarded the *Dátil* as a luxury. As I write there comes vividly to mind a chilly, mid-August morning in the Arizona plateau country, when two Navajo shepherdesses left their straggling flock to share in the warmth of our camp fire and pass the time of day. As they squatted by the flame, I noticed that one slipped some objects from her blanket into the hot ashes, but with such deft secretiveness that my eyes failed to detect what they were. Later as the woman rose to go, she raked away the ashes with a stick and drew out several blackened Yucca pods, which had been roasting while we talked. I can testify to the entire palatability of this cooked fruit (the rind being first removed), finding it pleasantly suggestive of sweet potato. Those fruits that morning were still green when plucked. Dr. H. H. Rusby informs me that the sliced pulp of the nearly ripe pods makes a pie almost indistinguishable from apple pie. The ripe fruit may be eaten raw, but the more usual custom among the Pueblo Indians, who would travel long miles in the pre-education days to gather the succulent, yellow pods and bring them home by the burro-load, was to cook them. Sometimes they were simply boiled, and on cooking the skin was removed, since it then separates easily from the pulp; but there was a more complicated process, resulting in a sort of conserve, that was considered better. This was to bake the fruit, peel it and remove the fibre, and then boil down the pulp to a firm paste. This was rolled out in sheets of about an inch in thickness, and carefully dried. Afterwards these were cut up into convenient sizes and laid away to be consumed either as a sweetmeat, or dissolved in water as a beverage, or employed like molasses on tortillas and bread.[5] The young flower buds of this and some other species of Yucca possess a considerable content of sugar and other nutritive principles, and by the aborigines are considered delicacies when cooked. Coville records a custom of the Panamint Indians who collected the swelling buds of the grotesque arborescent Yucca of the Mojave Desert known as the Joshua tree (*Yucca brevifolia,* Engelm.) and roasted them over hot coals, eating them afterwards either hot or cold.

The Yuccas have been useful to the desert people in other ways than as food, and we shall hear of them again in subsequent chapters. It is not remarkable, therefore, that the plant is imbued with sacred significance and enters in many ways into native religious ceremonies. Among the Navajos, *Yucca baccata* is called *hoskawn* and allusions to it are of frequent occurrence in the folk lore of that interesting race. Its leaves are the material out of which the ceremonial masks employed in the religious rites of these people are made. The Government has given particular distinction to this plant

[5] Bandelier, quoted by Harrington in "Ethnobotany of the Tewa Indians," Bull. 55, Bur. Amer. Ethnology.

by bestowing its Spanish name on the "Datil National Forest" of New Mexico.

The Cactus family, those especial plant children of the desert, yield some quite choice fruits, though they make us work to get them, hedged about as they are with vicious spines and bristles. Of several genera indigenous to the United States producing edible berries, the most widely distributed is *Opuntia*, embracing two quite different looking divisions, one with broad, flattened joints (the *Platopuntias*) and one with cylindric, cane-like joints (the *Cylindropuntias*). The former division includes the well-known Prickly Pears or Indian Figs, of which two species (*Opuntia vulgaris*, Mill., and *O. Rafinesquii*, Engelm.) occur in sandy or sterile soil of the Atlantic seaboard. Their seedy, lean, insipid berries, each an inch or so long, are edible in a way, but they are not at all in the same class with the fat, juicy "pears" of many of the species growing wild in the Southwestern desert country, where the genus is best represented. Even there, there is great choice in the fruits of different species, those of the broad-jointed sort being much the best. Such plants are called *nopal* by the Spanish-speaking Southwesterners and the fruit *tuna*. Among these *Opuntia laevis*, Coult., and the varieties of *O. Engel-*

occidentalis, Engl. and B. (the last abundant in Southern California) are especially valued. Better than these, however, are certain species introduced a century or more ago by the Franciscan Missionaries from Mexico, the motherland of the cacti. These are *O. megacantha*, Salm., and *O. Ficus-Indica*, Mill., and they now grow wild in many parts of California, especially about the old Mission towns, the fruit being annually harvested by the Mexican population. (See illustration facing page 18.)

The gatherer of tunas is faced by two difficulties—the rigid, needle-like spines that bristle on all sides of the plant, and the small tufts of tiny spicules that stud the fruit itself. The latter are really the more dangerous, because a touch transfers them from the tuna to the picker's flesh, there to stick and prick wickedly. If they happen to get into the mouth or upon the tongue, the pain is persistent and agonizing. With care, however, nothing of that sort need happen. Armed with a fork and a sharp knife, you spear your tuna firmly with the fork, give it a wrench and complete the parting from the stem by a slash of the knife. The next step is to peel the "pear," which is made up of a pulpy, seedy heart enveloped in an inedible rind. This may be readily got rid of in the following way: Handling the tuna with a

Gathering tunas, fruit of the nopal cactus, California.

glove or speared upon a fork, lay it upon a clean board, and holding it down slice off each end; then make a longitudinal cut through the rind from end to end; lay open both flaps of the rind, which may then be pressed back, separating along natural lines from the pulp. Indians and Mexicans dust the spicules off with a little brush of freshly gathered twigs, as of the California Sage. Such brushes they call *limpia-tunas*. (See illustration, page 174.)

Eaten raw, tunas of the better sort are refreshing and agreeable to most people, though the bony seeds are an annoyance unless one swallows them whole, after the Mexican fashion. The taste differs somewhat with the species, those that I have eaten possessing a flavor suggesting watermelon. The sugar content is considerable, and a very good syrup may be obtained by boiling the peeled fruits until soft enough to strain out the seeds; after which the juice may be boiled down further. No sugar need be added, unless a very sweet syrup is needed. Care should be exercised to select fruit that is really ripe; in some sorts maturity is slow to follow coloration. After all, though, it is Mexico where tuna raising and consumption have become an art, and the tuna market is an interesting feature in many Mexican towns. During the time of the harvest whole families go to the hills and camp out in the *Nopaleros* (the areas where the cactus grows) and live practically upon tunas alone. Mr. David Griffiths, in his monograph "The Tuna as a Food for Man,"[6] states that at such times about two hundred tunas a day constitute the ration of one individual. Large quantities are dried for future use and several products are also manufactured from the fresh fruit. One of these, called *queso de tuna* (that is, "tuna cheese"), is an article of sale in the Mexican quarters of our Southwestern towns. It is made by reducing the seeded tuna pulps to an evaporated paste, and is sent to market in the shape of small cheeses, dark red or almost black.

Another member of the Cactus family that is an important food source in the Southwest is the Sahuaro (*Cereus giganteus,* Engelm.). It is Arizona's floral emblem, and abounds throughout the southwestern part of that State and across the frontier into northern Mexico, forming at times in the desert strange, thin forests casting attenuated shafts of shade. It is one of the world's botanical marvels, a leafless tree with fluted, columnar trunk and scanty, vertical branches, rising sometimes to

[6] Bull. 116 Bur. Plant Industry, U. S. Dept. Agriculture.

the height of sixty feet and tipped in spring with numerous creamy, pink flowers. The fruit commonly goes by its Mexican name, *pitahaya.* It ripens in June and July, and somewhat resembles the tuna in form, with a juicy, seedy, crimson pulp. To civilized tastes, the fresh fruit is rather mawkish, less sweet than that of the related *pitahaya dulce,* which is common on the Mexican side of the border and is borne by *Cereus Thurberi,* Engelm. Nevertheless the Arizona pitahaya is of considerable food value and highly relished by the Indians of the region, particularly the older generation of Papagos, who make a festival of the opening of the pitahaya harvest, dating their new year from that event, and used to intoxicate themselves as a religious duty upon a sort of wine that they made for the occasion from the fermented first fruits.

The pitahayas are gathered with a twenty-foot pole, made of the rod-like ribs of some dead sahuaro lashed together and having a hook affixed to the tip, with which the fruit is dislodged. Such part of the crop as is not consumed raw is boiled down, as in the case of the tuna, the seeds removed, and then boiled again until the mass is reduced to a syrup. This is of a clear, light brown color, and pleasantly sweet, making a fair substitute for molasses and correspondingly good on bread or corn cakes. It is set away for winter consumption.[7] The inner part of the pitahaya may also be sun-dried, and will then keep for a long time. Sahuaro seeds are quite oily, and I am told by Mr. E. H. Davis that the Papagos dry them and grind them into an oleaginous paste, which they spread like butter on their tortillas. The ribs of this most useful plant are also employed by these same Indians as the basis of their stick-and-mud houses—a practice doubtless inherited from the ancients, as in many old cliff dwellings sahuaro ribs are found reinforcing adobe.

Our native Palms also have fruits to offer to the unexacting. On the Colorado Desert of South-eastern California, there is indigenous a stately palm known as the California Fan Palm (*Washingtonia filifera,* Wendl., var. *robusta*), which has been widely introduced into cultivation in the Southwest. In the cañons of the San Jacinto Mountains opening to the desert and in the desert foothills of the San Bernardino Mountains, as well as here and there in certain alkaline oases of the desert itself, extensive groves of this noble palm flourish—the remnant, it is

[7] For an interesting and detailed account of the Arizona Sahuaro harvest and uses, see Mr. Carl Lumholtz's "New Trails in Mexico."

California Fan Palm (Washingtonia), which furnishes food, textile and building materials.

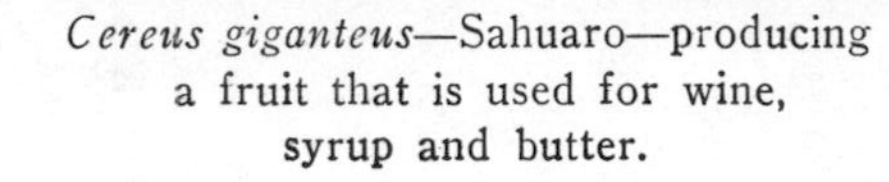

Cereus giganteus—Sahuaro—producing a fruit that is used for wine, syrup and butter.

believed, of far greater forests that probably existed in that region in primeval times. The mature fruit of the Washingtonia is berry-like and black, resembling a small grape or cherry, and is borne in huge compound clusters, which hang below the leafy crown of the tree in autumn and early winter. The relatively large seed is embedded in a thin pulp of sweetish flavor, which is edible, though it requires industry and a long pole to reach the fruit. These requisites were possessed by the old-time desert Indians, who used to make of the palm-berries an important feature in their diet, not only consuming the pulp both fresh and dried, but also grinding the seeds into a meal, which Dr. Edward Palmer thought as good as cocoanut. Similarly, of the Florida Palmetto berries the Seminoles made a flour and a syrup to sweeten it.

It is rather surprising to learn that the Toyon or California Holly (*Photinia arbutifolia*, Lindl.), universally used in California for Christmas decoration, was esteemed by the Indians for its edible berries. These when red ripe were eaten either raw out of hand or toasted by tossing in a basket with hot pebbles. Spanish Californians, in a simpler age than this, relished them too, steamed or parboiled, and served with sugar if there were any.

CHAPTER VI

WILD PLANTS WITH EDIBLE STEMS AND LEAVES

I often gathered wholesome herbs, which I boiled, or eat as salads with my bread.

Gulliver's Travels.

WHAT would you say to a dish of ferns on toast? It is quite feasible in the spring, if the Common Bracken (*Pteris aquilina,* L.) grows in your neighborhood—that coarse, weedy-looking fern with long, cord-like creeping root-stocks and great, triangular fronds topping stalks one to two feet high or more, frequent in dry, open woods and in old fields throughout the United States—the most abundant of ferns. The part to be used for this purpose is the upper portion of the young shoot, cut at the period when the fern shoot has recently put up and is beginning to uncurl. The lower part of the shoot, which is woody, and the leafy tip, which is unpleasantly hairy, are rejected. It is the intermediate portion that is chosen, and though this is loosely invested with hairs, these are easily brushed off. Then the cutting, which resembles an attenuated asparagus stalk, is ready for the pot. Divided into short lengths and cooked in salted, boiling water until quite tender—a process that usually requires a half to three quarters of an hour—the fern may be served like asparagus, as a straight vegetable, or on toast with drawn butter, or as a salad with French dressing. The cooked fern has a taste quite its own, with a suggestion of almond. Its food value, according to some experiments made a few years ago by the Washington State University, is reckoned as about that of cabbage, and rather more than either asparagus or tomatoes. Furthermore, the rootstocks of this fern are edible, according to Indian standards, and are doubtless of some nutritive worth as they are starchy, but the

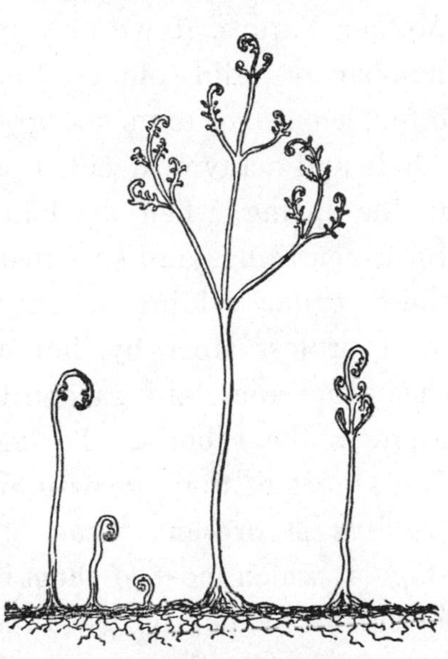

BRACKEN SHOOTS
(Pteris aquilina)

flavor does not readily commend itself to cultivated palates.

Dietitians who insist on the value of salads as part of a rightly balanced ration have a strong backer in Mother Nature, if we may take as a hint the large number of wild plants which everywhere freely offer themselves to us as "greens"—all wholesomely edible and many of decided palatability. Especially in the spring, when the human system is starving for green things and succulent, the earth teems with these tender wilding shoots that our ancestors set more or less store by, but which in these days of cheap and abundant garden lettuce and spinach we leave to the rabbits. To know such plants in the first stages of their growth, when neither flower nor fruitage is present to assist in identification—the stage at which most of them must be picked to serve as salads or pot herbs—presupposes an all-round acquaintance with them, so that the collector must needs be a bit of an expert in his line, or have a friend who is.

There is one, however, that is familiar to everybody—the ubiquitous Dandelion, whose young plants are utilized as pot-herbs particularly by immigrants from over sea as yet too little Americanized to have lost their thrifty Old World ways. It is a pleasant sight of spring days to see these new-fledged Americans dotting the fields and waste lots near our big cities, armed with knives, snipping and transferring to sack or basket the tender new leaves of the well-beloved plant, which, like themselves, is a translated European. The leaves are best when boiled in two waters to remove the bitterness resident in them; and then, served like spinach or beet-tops, they are good enough for any table. Old Peter Kalm, who has ever an eye watchful for the uses to which people put the wild plants, tells us the French Canadians in his day did not use the leaves of the Dandelion, but the roots, digging these in the spring, cutting them and preparing them as a bitter salad.

Then there is Chicory, which has run wild in settled parts of the eastern United States and to some extent on the Pacific coast, adorning the roadsides in summer with its charming blue flowers of half a day. Its young leaves, if prepared in the same way as those of the Dandelion, are relished by some. Preferably, though, the leaves are blanched and eaten raw as a salad. The blanching may be done in several ways. The outer leaves may be drawn up and tied so as to protect the inner foliage from the light and thus whiten it, or flower-pots may be capped over the plants. Another method is this:

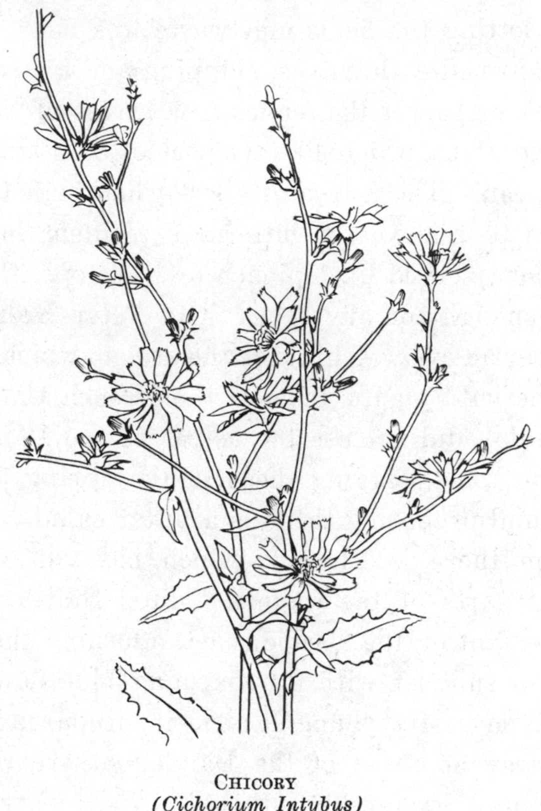

CHICORY
(Cichorium Intybus)

Dig up the roots in the autumn, cut back the tops to within an inch of the root-crown and bury the roots to within an inch of the top in a bed of loose mellow earth in a warm cellar. In a month or two, new leaves should appear, crisp and white and ready for the salad bowl.

Another old-fashioned pot-herb that may be gathered freely in the spring is the early growth of that familiar weed of gardens and waste places throughout the land, the homely Pigweed (*Chenopodium album,* L.), or Lamb's quarters. This latter queer name, by the way, like the plant itself, is a waif from England, and according to Prior[1] is a corruption of "Lammas quarter," an ancient festival in the English calendar with which a kindred plant (*Atriplex patula*), of identical popular name and usage, had some association. Of equal or perhaps greater vogue are the young spring shoots of the Pokeweed (*Phytolacca decandra,* L.) boiled in two waters (and in the second with a bit of fat pork) and served with a dash of vinegar. So, too, the first, tender sprouts of the common eastern Milkweed (*Asclepias Syriaca,* L.) have garnished country tables in the spring as a cooked vegetable, but the older stems are too acrid and milky for use. Mr. J. M. Bates, writing in "The American Botanist," speaks of this and of the closely related species, *A. speciosa,* Torr., of the region west of the Mississippi, as the best of all wild greens, provided they are

[1] "On the Popular Names of British Plants," R. C. A. Prior, M. D.

MILKWEED
(Asclepias Syriaca)

picked while young enough, that is, like asparagus sprouts and while the stems will still snap when bent. Young leaves and all are good in that stage of growth.

The Buckwheat family, which has yielded to civilization not only the grain that bears the family name but also the succulent vegetable Rhubarb, has some wild members with modest pretensions to usefulness. That common weed, naturalized from Europe, the Curled Dock (*Rumex crispus*, L.), for instance, is of this tribe; and its spring suit of radical leaves stands well with bucolic connoisseurs in greens. Another Rumex (*R. hymenosepalus*, Torr.), common on the dry plains and deserts of the Southwest and becoming very showy when its ample panicles of dull crimson flowers and seed-vessels are set, is famous there as a satisfactory substitute for rhubarb, which, indeed, the plant somewhat resembles. The large leaves, nearly a foot long, are narrowed to a thick, fleshy footstalk, which is crisp, juicy and tart. These stalks, stripped off before the toughness of age has come upon them, and cooked like rhubarb, are hardly distinguishable from it. Westerners know it as Wild Rhubarb, Wild Pie Plant, and Cañaigre. Under the last name it has some celebrity as tanning material, the tuberous roots being rich

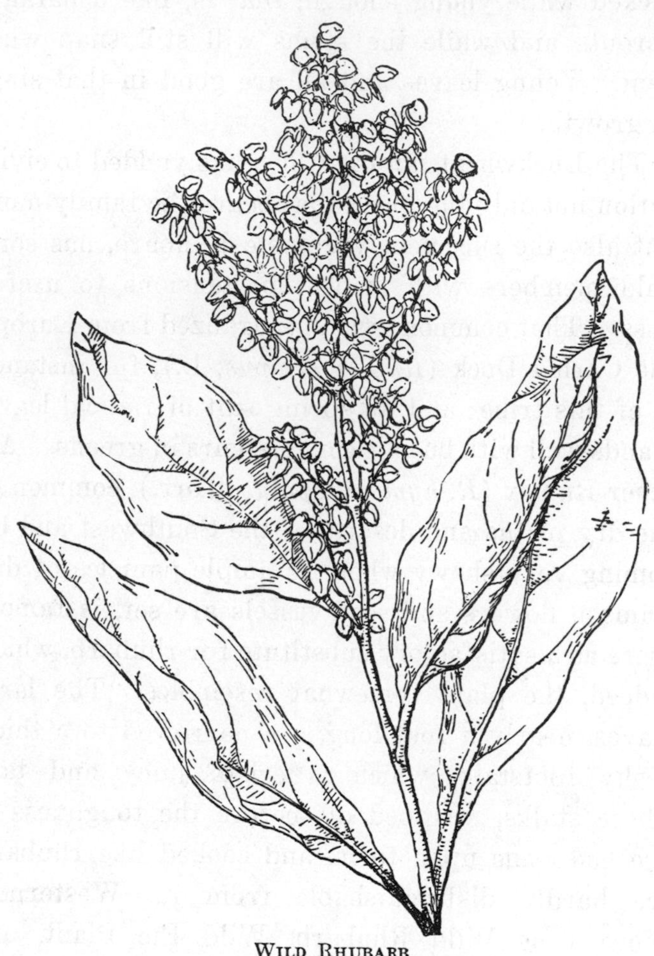

WILD RHUBARB
(Rumex hymenosepalus)

in tannin and having been long used by the Indians in treating skins. The tannin is extracted by leaching the dried and ground roots.

To the same family belongs the vast western genus *Eriogonum,* which includes that famous honey plant of the Pacific coast known as Wild Buckwheat. Some members of this genus are prized by the Indians and children for the refreshing acidity of the young stems—a quality of distinct value in the arid regions where many of them grow and where one is "a long way from a lemon." Among such is *Eriogonum inflatum,* T. & F., the so-called "Desert Trumpet" or "Pickles," found abundantly on the southwestern desert as far north as Utah and eastward to New Mexico. It is remarkable for its bluish-green, leafless stalks, hollow and puffed out like a trumpet, sometimes to the diameter of an inch or so, and rising out of a radical cluster of small heart-shaped leaves. The stems before flowering are tender and are eaten raw.

The peppery, anti-scorbutic juices of the Mustard family supply a valuable element in the human dietary everywhere; and besides the important vegetables and condiments that represent it in our gardens—such as cabbage, turnips, radishes, horse-radish, etc.—there are several species growing wild that have been proved of worth. Water-cress, known to everybody (*Nasturtium officinale,* R. Br.) and originally introduced, at least in the East, from Europe, is now a common aquatic throughout a large part of the United States and Canada. The waters of springs and brooks are often found thickly blanketed with green coverlets of this plant dotted with the tiny white flowers, and lending spice to the wayfarer's luncheon. Winter Cress, Yellow Rocket, or Barbara's Cress (*Barbarea vulgaris,* R. Br.) used to be very generally eaten by people of humble gastronomic aspirations, so that it has acquired the additional name of Poor Man's Cabbage, being prepared either as a pot-herb or as a salad. It is abundant by roadsides and in low-lying fields quite across the continent, and, in fact, almost around the world, and was no doubt cultivated in our colonial gardens. Even in winter, when the snow melts enough to show bare patches of earth, the tufted, thickish leaves of this sturdy mustard are frequently revealed, green and alive, hugging the ground. The lower leaves are of the shape that botanists call lyrate—that is, long and deeply lobed, with one to four pairs of segments and a terminal one large and roundish. In early spring it sends up a spike of showy, yellow, four-petaled flowers. Quite similar

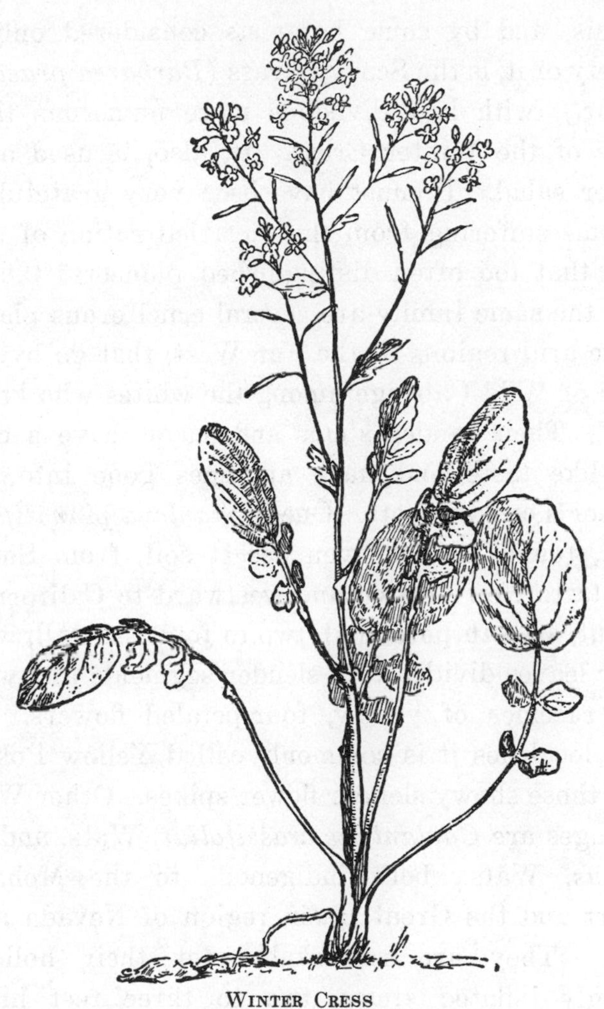

WINTER CRESS
(Barbarea vulgaris)

to this, and by some botanists considered only a variety of it, is the Scurvy Grass (*Barbarea praecox,* R. Br.), with leaf divisions more numerous than those of the Winter Cress. It, also, is used as a winter salad. It must have been very grateful to systems suffering from the unvaried ration of salt meat that too often distinguished pioneers' tables.

Of the same family are several cruciferous plants of the arid regions of the Far West, that go by the name of Wild Cabbage among the whites who know them. Their tender stems and leaves have a cabbage-like taste and have at times gone into the pioneer's cooking pots. One is *Stanleya pinnatifida,* Nutt., found in dry, even desert soil, from South Dakota to New Mexico and westward to California, a stout, smooth perennial, two to four feet tall, with lower leaves divided into slender segments and with long racemes of yellow, four-petaled flowers. In some localities it is commonly called Yellow Poker, from those showy slender flower spikes. Other Wild Cabbages are *Caulanthus crassifolius,* Wats. and *C. inflatus,* Wats., both indigenous to the Mohave Desert and the Great Basin region of Nevada and Utah. They are remarkable for their hollow, strongly inflated stems, two to three feet high. These, bearing racemes of small, deep-purple flowers with curious wavy, crisped petals, make the plants easily recognized. The young plants, while still tender, are edible but need to be cooked. The process pursued by the Panamint Indians is thus described by Coville: "The leaves and young stems are gathered and thrown into boiling water for a few minutes, then taken out, washed in cold water, and squeezed. The operation of washing is repeated five or six times, and the leaves are finally dried, ready to be used as boiled cabbage. Washing removes the bitter taste and certain substances that would be likely to produce nausea or diarrhoea."

One would suppose that the stinging Nettle (*Urtica dioica*, L.) would be as unlikely a subject as one could readily find to supply a morsel wherewith to tickle the palate. Nevertheless, this "naturalized nuisance," as good old Doctor Darlington of "Flora Cestrica" fame testily styles it, has long been valued as a vegetable in Europe, whence the plant has come to us. There the tender shoots, cut before the flowering stage, were served in old times on the tables of the well-to-do as well as of the peasantry. On a day in February, 1661, Mr. Samuel Pepys, of immortal memory, ingenuously set down in his diary the fact that calling upon one

Mr. Simons in London, he found the gentleman abroad, "but she, like a good lady, within, and there we did eat some nettle porridge, which was made on purpose to-day for some of their coming, and was very good." Was it not Goldsmith who wrote that a French cook of the olden time could make seven different dishes out of a nettle-top?

Along our Southwestern border from Texas to California and southward into Mexico a species of Amaranth grows (*Amaranthus Palmeri,* Wats.), known to the Mexicans and Indians as *quelite* (a general name among the Mexican population, I believe, for greens) or more specifically as *bledo.* The latter word is good Spanish for "blite," an Old World pot-herb. *Quelite* is highly regarded when young and tender as a vegetable for men, and, when cut and stacked, as a winter feed for cattle. It is a stout, weedy annual, two to four feet high, the ovate leaves one to four inches long on footstalks about twice that length, the greenish flowers of two sexes (on different plants) disposed in long, dense chaffy spikes. Only the young plants should be gathered; they should then be boiled without delay, and the result, in the judgment of white people who know it, is a dish resembling asparagus in flavor, and rather superior to spinach. Mexicans and Indians have used it extensively, as well as other species of the genus. To Americans Amaranths are just Pigweed.

This little course in wild pot-herbs may now be closed with mention of three members of the Portulaca family. These plants are marked by smooth, succulent, thickish leaves, and though humble herbs, they are usually found, when found at all, in sufficient abundance to be very noticeable. Most familiar is the little prostrate plant common everywhere in fields and waste places, called Purslane (*Portulaca oleracea,* L.). It is generally regarded by Americans as a weed and provokes the temper by its stubborn persistence in turning up after it has apparently been eradicated. It has, however, held quite a respectable social position abroad, where gardeners have cultivated it and developed it as a wholesome vegetable useful not only as a pot-herb but for salads and pickles.[2] On the Pacific slope a cousin of the Purslane, known as Miner's or Indian Lettuce (*Montia perfoliata,* Howell), is abundant in shady places. It may be recognized by its dainty racemes of tiny white flowers beneath which a pair of opposite leaves united at their bases forms a cup or saucer around the stem, a diagnostic feature of the plant. The Indians were very fond of the pleasant

[2] Eaten raw it is a valuable anti-scorbutic.

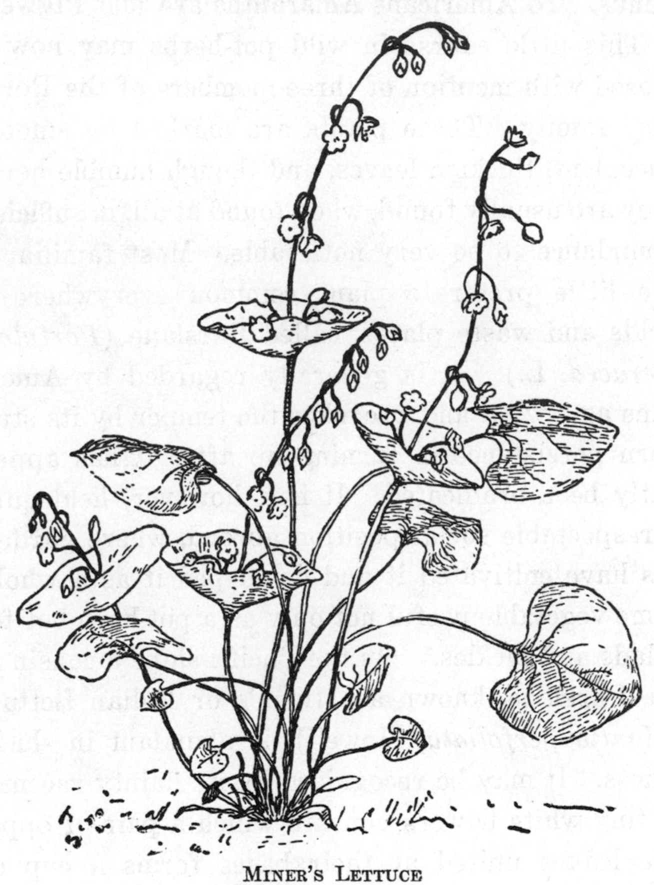

MINER'S LETTUCE
(Montia perfoliata)

succulence of the stem and leaves and their consumption of the herb led the white pioneers to try it. It makes, indeed, a palatable enough dish, either raw with a sprinkling of salad dressing or boiled and served like spinach. Stephen Powers tells of a certain tribe of California Indians who were accustomed to lay the leaves near the nests of red ants, which running over the greens would flavor them with a formic acidity that served in lieu of vinegar![3] A Spanish Californian name for the plant is *petota.* A famous old time salad consists of a peeled *tuna* or prickly pear, nestling amid *petota* leaves, served with a dressing of olive oil, salt, pepper and vinegar.

Also of California is another of the Portulaca kin, the crimson flowered *Calandrinia Menziesii,* Gray, the familiar Red Maids which spreads carpets of glowing color in spring fields and vineyards. Its juicy herbage is useful as a garnish or pot-herb; but some Indian tribes found a greater value in the small, shining black seeds, resembling grains of gunpowder, which they ground for pinole. And there is the Ice-plant (*Mesembryanthemum crystallinum,* L.) of southern California beaches. The succulent leaves, frostily beaded, serve handily for picnickers' salads.

If, as we have seen, the Nettle may be made to

[3] "Contributions to North American Ethnology," vol. III, 425.

grace the table, it is quite credible that within the spiny armor of the Cactus tribe nutrition may be hiding. As a matter of fact, in the Southwest the Mexican and Indian population resort to the Nopal (that is, the flat-jointed sort of *Opuntia*) not only for the tuna fruit, as described in a previous chapter, but also for the succulent flesh of the stem, which may be made to do duty as a vegetable. The Mexicans call these flattened joints *pencas,* and gather the young ones when about half grown and before the spines have hardened. Cut into narrow strips, boiled until tender and served with a tasty dressing or just salt and pepper, they are about in the class of string beans, particularly grateful to desert dwellers whose craving for green food it is not always easy to satisfy. There is a bluish-green, procumbent cactus without spines (*Opuntia basilaris,* Engelm.) common in the southwestern deserts, that has been in particular favor with the Indians, and the Panamint method of preparing it, as recorded by Mr. Coville,[4] may be stated here: In May or early June the fleshy joints of the season's growth, as well as the buds, blossoms and immature fruit, are distended with sweet sap. The joints are then broken off and collected, carefully rubbed with grass to remove the

[4] The American Anthropologist, October, 1892.

tiny bristles, and spread in the sun to dry. After being thoroughly dried, they will keep indefinitely, and are boiled as required and eaten with a seasoning of salt. An alternative process is to steam the joints for about twelve hours in stone-lined pits first made hot by a fire of brush. The cactus, thus cooked, may be eaten at once or dried and laid away for future use. It then has the texture and appearance of unpeeled dried peaches.

From the curious, cylindrical, keg-like bodies of another cactus of the Southwest (*Echinocactus* sp.), termed *bisnaga* by the Mexicans, or Barrel Cactus by polite Americans (others sometimes style it Nigger-head), a sort of conserve used to be made by the Papago Indians of Arizona—the prototype of the so-called "Cactus Candy" of city shops. The process, as described by Dr. Edward Palmer, was to pare away the thorny rind of a large specimen and let it remain several days "to bleed." Then the pulp was cut up into pieces of suitable size and boiled in the syrup of the Sahuaro *pitahayas,* obtained as described in the preceding chapter. Another and more important use of this cactus will be described later.

Few plants of the Southwestern desert region are more interesting and useful than the Agave, a genus of the Amaryllis family. Its general aspects are made familiar through the well-known Century Plant of cultivation. There are a dozen species or more indigenous within the limits of the United States, ranging mostly along the Mexican border from Texas to California. For years—ten to twenty, it may be—the plant devotes itself exclusively to developing a rosette of slender, pulpy, dagger-pointed leaves, stiff and fibrous. Then some spring day, within the center of this savage leaf-cradle, a conical bud is born and develops quickly, a foot a day it may be, into a huge, asparagus-like stalk, twelve or fifteen feet tall, that breaks out at the summit into clusters of yellow blossoms. This long delayed consummation costs the plant its life, and with the maturing of its seeds it turns brown and withers away. It is from a Mexican species of Agave that the Mexicans manufacture their desolating drinks *pulque* and *mescal*. The United States species, however, have been little turned to such account, but as a nutritive food source they have from very ancient times been important to the Indians. This food shares with the fiery Mexican drink the name *mescal*. Even at the present day, when the ease of extracting a meal from a tin can has been the cause of relegating many an honest old-time cookery to oblivion, there are Indians who pack up every spring and repair to the mescal fields, there to open again the ancient baking pits which their fathers and their fathers before them had used, and camp for a week at a time, cutting and cooking, feasting and singing, and telling once more the immemorial legends of their race.

The process of preparing mescal as I happen to have observed it in California is this: The succulent, budding flower-stalks when just emerging from amid the leaves are cut out with an axe, or better yet with a native implement fashioned for the purpose—a long, stout lever of hard wood (oak or mountain mahogany) beveled at one end like a chisel. They are then trimmed of their tips and all adhering leafage, the desirable portion being the butt, which is filled with all the pent-up energy that the plant was holding in reserve for the supreme act of flower and seed production. Meantime, a circular pit, about a foot and a half deep and five or six feet in diameter, has been prepared—usually one that has been used in previous years being dug out. This is lined side and bottom with flat stones, and a huge fire of dry brush started in it, care being taken to use no wood that is bitter. When the fire has burned down, the mescal butts are placed in the hot ashes, covered over with more hot ashes and heated stones from the sides of the pit, and all is then buried beneath a mound of earth. There the mescal is left to steam until some time the next day, like the four-and-twenty blackbirds of the nursery rhyme in their pie. When the pit is opened the mescal, still hot and now charred on the outside, is drawn out, the burnt exterior pared off, and the brown, sticky inside laid bare, to be eaten on the spot or laid away to cool and be transported home for future use. If the buds have been cut young enough, mescal is tender and sweet, the flavor suggesting a cross between pine-apple and banana and pleasant to most white palates. Indians are extravagantly fond of it, and it is rare indeed that the stock carried home lasts over the following summer. Should the buds be too old when cooked, the result is unpleasantly fibrous, though in such cases one need only chew until the edible part is consumed, when the fibre may be spat out. Mr. Coville, in his account of the Panamints above quoted, speaks of finding at some forsaken Indian camps along the Colorado River, dried and weathered wads of chewed mescal fibre—visible reminders of forgotten feasts.

Denizens of the same region with the Agaves, and

Southwestern Indian cutting mescal (*Agave deserti*) for baking.

somewhat resembling them, are several species of Dasylirion, but the leaves, which form a crown upon a central stem, are much narrower and the small flowers are white and constructed on the plan of the lily. They are called, in popular parlance, Bear-grass, from Bruin's fondness for the tender stalks, or more generally by their Mexican name, *Sotol.* The budding flower-stalks are to some extent used like mescal—roasted and eaten. So, too, the beautiful *Yucca Whipplei,* Torr., abundant throughout Southern California and adjacent regions, has been made to add variety to the aboriginal menu. The splendid flower masses of this plant, several feet in length and rising in pure white spires out of a bristling clump of slender, rigid, spine-tipped leaves, are a famous sight in parts of the Southwest. Americans call this Yucca "Spanish Bayonet," or sometimes more poetically "The Lord's Candle." To Mexicans it is *quiote,* one of the many Aztec terms that survive with little mutilation in the Spanish dialect of the Southwest. The flower-stalk, when full grown but before the buds expand, is filled with sap and is edible, cut into sections and either boiled or roasted in the ashes. The tough rind should first be peeled off. The flower buds, too, make a palatable vegetable, if boiled, and serve as a succulent side-dish to the camper's usually monotonous dry diet.

On the Southeastern rim of our country from North Carolina to Florida, a common tree is the Cabbage Palmetto (*Sabal Palmetto,* R. & S.), which South Carolina has adopted as so peculiarly her own that she is known as the Palmetto State. It is a palm of much the general look of the California Fan Palm, though it never attains so great a height as the latter often does. All palms grow by the development of a central, terminal leaf-bud, and this in some species—the Palmetto is one—is turned to account as an edible, being popularly known as a "cabbage." When cooked, the Palmetto cabbage is a tender, succulent vegetable, though the harvesting of the buds is a wasteful practice, unless it is desired to clear the land, as cutting them out kills the trees.

We have it on the authority of Holy Writ that Nebuchadnezzar, king of Babylon, foregathered for a season with the beasts of the field and ate grass as oxen, finding it, it is to be assumed, a sustaining ration. The Indians of California, curiously enough, long ago acquired and maintained more persistently than the royal Babylonian a similar habit of turning themselves out to pasture, to feast upon the patches of wild clover. This they ate raw and with greedy avidity, before the flowering stage, while the plants were still young and tender. In fact, clover was another of the aboriginal food plants esteemed as so important as to be honored with especial dance ceremonies. Chesnut speaks of seeing groups of Indians in Mendocino County, California, wallowing in the wild clover, plucking the herbage and eating it by the handful. Its nutritive content is unquestioned, if only one have the digestive organs to handle it, chemical analysis of the leaves showing the presence of food elements in good degree. Intemperate indulgence, however, is liable to cause bloat and severe indigestion. The Indians, to obviate this, learned that dipping the leaves in salted water, or munching with them the parched kernels of the Pepper-nut (the fruit of the California Laurel, *Umbellularia Californica*) is efficacious.[5] Not all species of clover are considered equally good. The favorite, still to quote Chesnut, is the so-called "sweet clover" (*Trifolium virescens*, Greene), distinguished by stout, succulent stems, ovate leaflets, large, inflated yellow and pink flowers,

[5] V. K. Chesnut, "Plants Used by the Indians of Mendocino Co., California."

and a noticeable sweetness of taste. Of this species even the flowers are eaten. Next to this in flavor is the "sour" or "salt clover" (*T. obtusiflorum*, Hook.), with narrow, saw-toothed leaflets, whitish blossoms with purple centers, and a clammy, acidulous exudation that covers the leaves and flowers.

I had thought to close this chapter here, when a correspondent who is a veteran camper, Dr. Robert T. Morris, of New York, reminds me of certain other plants which he has found so useful that I add them. The Spotted Touch-me-not or Jewel-weed (*Impatiens fulva*, Nutt.) he has depended upon for weeks at a time in the northeastern wilderness, where, under the name of Lamb's-quarters, it is commonly regarded as an important vegetable food. It luxuriates beside shady rills, and its orange-colored spotted flowers, followed by fat pods that burst at a touch, are familiar to all. Excellent, too, in early spring, are the latent buds of the Cinnamon and Interrupted Ferns (*Osmundas*), rivals of the chestnut in flavor and size. Then those leathery lichens common on rocks and known as Rock-tripe (*Umbilicaria*), so often included in the menus of old-time hunters and *voyageurs*, have value. "They make," to quote Dr. Morris, "an excellent pottage, although the addition of a little bacon or deer meat or wild onion improves the flavor very much."

CHAPTER VII

BEVERAGE PLANTS OF FIELD AND WOOD

And sip with nymphs their elemental tea.
Pope.

MAN dearly loves a sup of drink with his meat, and when our pioneer ancestors in the American wilderness ran short of tea and coffee and craved a change from cold water, they found material for more or less acceptable substitutes in numerous wild plants. Particularly during the American Revolution was interest awakened in these, and several popular plant-names still current date from those days of privation. Again during our Civil War the attention of residents in the South was similarly drawn to the wild offerings of nature. A literary curiosity, now rare, of those dark days may still be turned up in libraries, a book entitled "Resources of Southern Fields and Forests . . . with practical information on the useful properties of the Trees, Plants and Shrubs," by Francis Peyre Porcher, Charleston, S. C., 1863, the writer being then a surgeon in the Confederate Army.

Among such beverage plants one of the best known is a little shrub, two or three feet high, frequent in dry woodlands and thickets of the eastern half of the continent from Canada to Texas and Florida, commonly called New Jersey Tea, the *Ceanothus Americanus,* L., of the botanists. It is characterized by pointed, ovate, toothed leaves, two or three inches long, strongly 3-nerved, and by a large, dark red root, astringent and capable of yielding a red dye. This last feature has given rise to another name for the plant in some localities—Red Root. In late spring and early summer the bushes are noticeable from the presence of abundant, feathery clusters of tiny, white, long-clawed flowers which, if examined closely, are seen to resemble minute hoods or bonnets extended at arm's length. The leaves contain a small proportion of a bitter alkaloid called ceanothine, and were long ago found to make a passable substitute for Chinese tea. During the Revolutionary War an infusion of the dried leaves as a beverage was in common use, both because of the odium attached to real tea after the taxation troubles with England, and from motives of necessity. Connoisseurs claim that the leaves should be dried in the shade. There are a score or more of species of *Ceanothus* indigenous to the Pacific coast, where

New Jersey Tea
(Ceanothus Americanus)

they are known as "myrtle" or "wild lilac"; but I have not heard of their leaves being used like those of the eastern species mentioned. These plants will be referred to again in the chapter on Vegetable Soaps.

Another of the Revolutionary War substitutes was the foliage of the so-called Labrador Tea (*Ledum Groenlandicum*, Oeder), a low evergreen shrub of cold bogs throughout Canada and the northeastern United States as far south as Pennsylvania. A distinguishing feature is in the narrow, leathery leaves with margins rolled back and a coating of rusty wool on the under side. When pinched the foliage exhales a slight fragrance.

The familiar Sassafras of rich woods, old fields and fencerows on the Atlantic side of the country attracted attention very early in colonial days, and all sorts of virtues as a remedial agent were ascribed to it. During the Civil War, Sassafras tea became a common substitute for the Chinese article, and as a spring drink for purifying the system it still has a hold on the popular affection. The root is the part generally utilized, an infusion of the bark being made which is aromatic and stimulant. The flowers also may be similarly treated.

Of the same family with the Sassafras and of

much the same distribution is the common Spicewood, Wild Allspice, or Feverbush[1] (*Lindera Benzoin*, Blume), a shrubby denizen of damp woods and moist grounds, easily recognized in early spring by the little bunches of honey-yellow flowers that stud the branches before the leaves appear. The whole bush is spicily fragrant, and a decoction of the twigs makes another pleasant substitute for tea, at one time particularly in vogue in the South. Dr. Porcher states that during the Civil War soldiers from the upper country in South Carolina serving in the company of which he was surgeon, came into camp fully supplied with Spicewood for making this fragrant, aromatic beverage. André Michaux, a French botanist who traveled afoot and horse-back through much of the eastern United States when it was still a wilderness, half starving by day and sleeping on a deer-skin at night, has left in his journal the following record of the virtues of Spicewood tea, served him at a pioneer's cabin: "I had supped the previous evening [February 9, 1796] on tea made from the shrub called Spicewood. A handful of young twigs or branches is set to boil and

[1] Also called Benjamin-bush, corrupted from benzoin, an aromatic gum of the Orient which, however, is derived from quite another family of plants. French-Canadians used to call the Spicewood, *poivrier*, which means pepper plant.

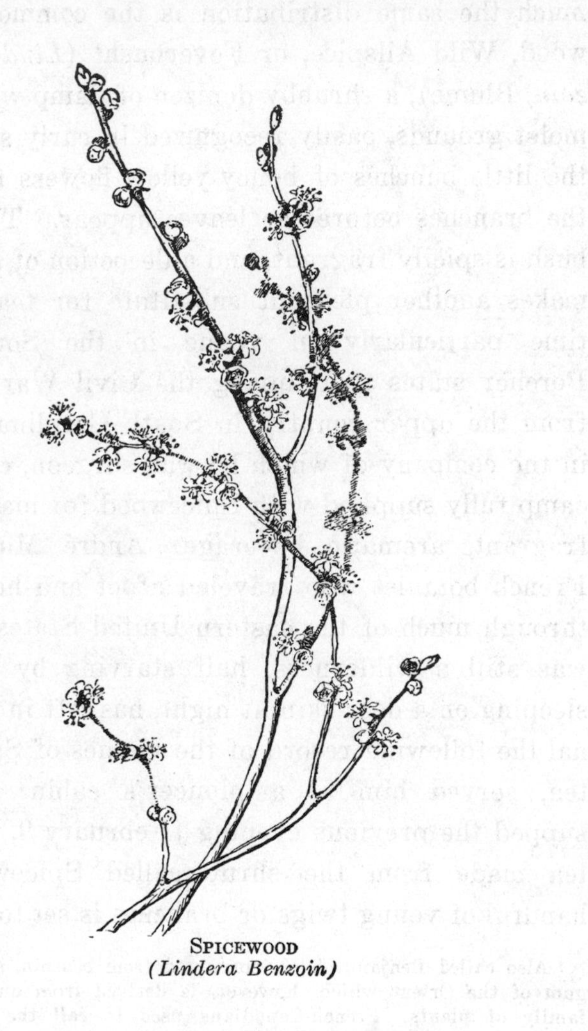

SPICEWOOD
(Lindera Benzoin)

after it has boiled at least a quarter of an hour, sugar is added and it is drunk like tea. . . . I was told milk makes it much more agreeable to the taste. This beverage restores strength, and it had that effect, for I was very tired when I arrived." The scarlet berries that cling like beads to the branches in the autumn used to be dried and powdered for use as a household spice, whence, obviously, the name Wild Allspice sometimes given to the shrub.

The warm, birchy flavor of the creeping Wintergreen (*Gaultheria procumbens*, L., the use of whose berries was noted in the previous chapter) could hardly have failed to attract attention to the plant as a likely substitute for Chinese tea when the latter was unobtainable; and one of its popular names, Teaberry, indicates that that is what happened—an infusion of the leaves being made. A pleasant and wholesome drink may also be made from the foliage of one of the Goldenrods—*Solidago odora*, Ait. This is a slender, low-growing species with one-sided panicles of flowers, not uncommon in dry or sandy soil from New England to Texas and distinguished by an anise-like fragrance given off by the minutely dotted leaves when bruised. A common name for it is Mountain Tea, and in some parts of the country the gathering of the leaves to dry and peddle in the winter has formed a minor rural industry, yielding a modest revenue.

The devotees of coffee, too, have found in the wilderness places substitutes for their cheering cup. One of these is the seed of the Kentucky Coffee-tree (*Gymnocladus Canadensis*, Lam.), a picturesque forest tree with double-compound leaves occurring from Canada to Oklahoma. In winter it is conspicuous because of the peculiar clubby bluntness of the bare branches, due to the absence of small twigs and branchlets, which gives to the whole tree a lifeless sort of look that gained for it among the French settlers the name *Chicot*, a stump. In the autumn the female trees (the species is diœcious) are seen hanging with brown, sickle-like pods six to eight inches long and an inch or two wide, and containing in the midst of a sweetish pulp several hard, flattish seeds. If we are to judge from the popular name it was probably the pioneers in Kentucky that first had an inspiration to roast these seeds and grind them for beverage purposes. The fact is, however, that a century ago such use of them was quite prevalent in what was then the western wilderness, and travelers' diaries of the time make frequent mention of the practice. The journal, for instance, of Major

Long's expedition to the Rocky Mountains in 1819-20 records that while in winter camp on the Missouri River near Council Bluffs, the party substituted these seeds for coffee and found the beverage both palatable and wholesome. Thomas Nuttall, the botanist, who botanized the following year around the mouth of the Ohio River, testifies to the agreeableness of the parched seeds as an article of diet, but thought that as a substitute for coffee they were "greatly inferior to cichorium."

Cichorium is the botanists' way of saying Chicory, the plant that has been referred to already as producing leaves useful as a salad. Its root has had a rather bad name as an adulterant of coffee, in which delusive form it has perhaps entered more human stomachs than the human mind is aware of. As a drink in itself, sailing under its own colors, Chicory is not a bad drink, the root being first roasted and ground. It is rather surprising, by the way, to learn that a palatable beverage is possible from steeping the needles of the Hemlock tree (*Tsuga Canadensis*, Carr.)—which is not to be confused with the poisonous herb that Socrates died of. Hemlock tea is, or at least used to be, a favorite drink of the eastern lumbermen, and I have myself drunk it with a certain relish. Similarly the leaves of the magnificent Douglas Spruce (*Pseudotsuga taxifolia*, Britt.) of the Pacific coast produce by infusion a beverage which many Indians and some whites have esteemed as a substitute for coffee.

The Mint family, well advertised by the pronounced and usually agreeable fragrances given off by its members, has been utilized as a source less of ordinary beverages than of medicinal teas, administered in fevers and digestive troubles. Such plants of the former sort as have come to my notice are all western. One of these has, in fact, played both rôles. This is the aromatic little vine known in California as Yerba Buena (the botanist's *Micromeria Douglasii*, Benth.), found in half shaded woods and damp ravines of the Coast Ranges from British Columbia to the neighborhood of Los Angeles. Its dried leaves steeped for a few minutes in hot water make a palatable beverage mildly stimulating to the digestion, and, like real tea, even provocative of gossip; for it is an historic little plant, this Yerba Buena, which gave name to the Mexican village out of which the city of San Francisco afterwards rose. The two words, which mean literally "good herb," are merely the Spanish for our term "garden mint," of whose qualities the

YEBBA BUENA
(Micromeria Douglasii)

wild plant somewhat partakes.[1] Of the Mint tribe, also, is the herb Chia, about whose edible seeds something has been said. At the present day, Chia is better known as a drink than as a food. A teaspoonful of the seeds steeped in a tumbler of cold water for a few minutes communicates a mucilaginous quality to the liquid. This may be drunk plain, but among the Mexicans, who are very fond of it as a refreshment, the customary mode of serving it is with the addition of a little sugar and a dash of lemon juice. The tiny seeds, which swim about in the mixture, should be swallowed also, and add nutrition to the beverage. A Spanish-California lady of the old school gave me my first glass of Chia, and recommended it as "*mejor que ice-cream*" (better than ice cream).

Of quite a different sort, but equally refreshing and easy to decoct, is the woodland drink called "Indian lemonade," made from the crimson, berry-like fruits of certain species of Sumac. East of the Rockies there are three species abundant, dis-

[1] The mint of the gardens (*Mentha viridis* and, to a less extent, *M. piperita*) is a common escape in damp ground and by streamsides throughout the country. In the Southwest the leaves, under the name of Yerba Buena, are used in the same way as those of *Micromeria*. A steaming hot infusion of mint leaves is a bracing beverage highly esteemed by tired, wet vaqueros coming in at evening from their day's work on the range.

SUMAC
(Rhus glabra)

tinguished by compact, terminal, cone-like panicles of white flowers and pinnate leaves that turn all glorious in the autumn in tones of orange and red. They are *Rhus typhina*, L. (Staghorn Sumac), *R. glabra,* L. (Smooth Sumac), and *R. copallina,* L. (Dwarf Sumac). The first is sometimes a small tree; the others are shrubs. In the Rocky Mountain region and westward *Rhus trilobata,* Nutt., is frequent—the Squaw-bush, as it is called, because the branches are extensively used by the Indian women in basketry; and on the Pacific coast, *Rhus ovata,* Wats., and *R. integrifolia*, B. & H., stout shrubs or small trees, occur. The last two have leathery, entire leaves quite unlike those of the eastern species, and the white or pinkish flowers are borne in tight little clusters. The berries of all these sumacs are crimson and clothed with a hairy stickiness that is pleasantly acid and communicates a lemon-like taste to water in which the fruit has been soaked for a few minutes. These plants—particularly the western species—are often found growing on hot, waterless hillsides, and their fruits offer a grateful refreshment to the thirsty traveler, whether sucked in the mouth until bared of their acid coating, or steeped in water to serve as a woodland lemonade. The three far western species are com-

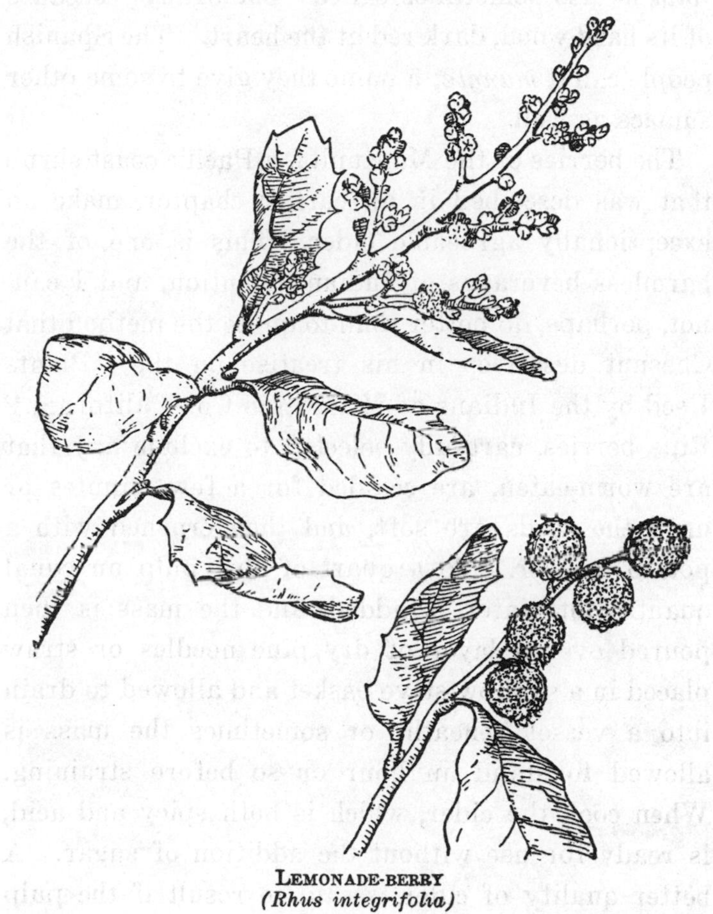

LEMONADE-BERRY
(Rhus integrifolia)

monly known as Lemonade-berry, and *R. integrifolia* is also sometimes called "mahogany" because of its hard wood, dark red at the heart. The Spanish people call it *mangle,* a name they give to some other sumacs as well.

The berries of the Manzanita, a Pacific coast shrub that was described in an earlier chapter, make an exceptionally agreeable cider. This is one of the harmless beverages of Indian invention, and I cannot, perhaps, do better than to quote the method that Chesnut describes in his treatise on the "Plants Used by the Indians of Mendocino Co., California." Ripe berries, carefully selected to exclude any that are worm-eaten, are scalded for a few minutes or until the seeds are soft, and then crushed with a potato masher. To a quart of this pulp an equal quantity of water is added, and the mass is then poured over a layer of dry pine needles or straw placed in a shallow sieve basket and allowed to drain into a vessel beneath; or sometimes the mass is allowed to stand an hour or so before straining. When cool, the cider, which is both spicy and acid, is ready for use without the addition of sugar. A better quality of cider is said to result if the pulp alone is used. The dried berries, in the latter case, are pounded to a coarse powder, and then by clever manipulation and tossing in a flat basket—a process at which the Indian woman is an adept—the heavier bits of seed are made to roll off while the fine particles of pulp cling to the basket.

The desert, too, has its beverage plants. There, if anywhere, pure water takes its place as the most luxurious of drinks, and the sands bear at least one group of plants from which good water may be obtained, namely, the Barrel Cactuses (*Echinocactus*) of the Southwest, of which something has been said under another head. The juices of most cacti, while often plentiful, are as often bitter to nauseousness; but those of the Barrel Cactus—or at least of certain species—are quite drinkable, and the rotund, keg-like plants serve a very important purpose as reservoirs of soft water. This is readily obtainable by horizontally slicing off the top and pounding up the succulent, melon-like pulp with a hatchet or piece of blunt, hard wood that is not bitter. In this way the watery content is released and may be dipped out with a cup. In the case of some species, I believe, the juice is too much impregnated with mineral substances to be drinkable; but in others—as *Echinocactus Wislizeni*, Engelm., *E. Emoryi*, Engelm., and *E. cylindraceus*, Engelm.—the fluid obtained is clear and pleasant to the taste, quenching the thirst satisfactorily. An odd and all but forgotten use of these vegetable water barrels of the desert is their former employment by Indians as cooking vessels. The fleshy interior was scooped out and the shell treated as a pot, into which water (secured by the mashing up of the pulp) was poured, heated with hot stones and these withdrawn as they cooled and replaced with hotter. Meantime the meat and other edibles were dropped in and allowed to simmer until done. Upon breaking camp, the cook abandoned her impromptu kettle, depending upon finding material for a new one at the next stopping place.

Throughout the arid and semi-desert regions of the Southwest from New Mexico to Southern California, a peculiar plant called *Ephedra* by the botanists is abundant. There are several recognized species but all have so strong a family resemblance that in popular parlance they are lumped as one and spoken of as Desert Tea or Teamster's Tea. They are shrubby plants, two or three feet high, greenish-yellow and distinguished by slim, cylindrical, many-jointed stems and abundant opposite branches, the leaves reduced to mere scales. The clustered flowers, inconspicuous and borne in the axils of the branches, are of two sorts on different plants, the pistillate producing solitary, black seeds of intense

Echinocactus, a vegetable water barrel of the Southwestern deserts.

A California Soap Root, *Chenopodium Californicum.* (See page 174.)

bitterness. The plant is well stocked with tannin, and an infusion of the branches—green or dried—in boiling water has long been in favor with the desert people, red and white. Desert Tea was first adopted by the white explorers and frontiersmen as a medicinal drink, supposed to act as a blood purifier and to be especially efficacious in the first stages of venereal diseases; but its use at meals as an ordinary hot beverage in substitution for tea or coffee is by no means uncommon, and cowboys will sometimes tell you they prefer it to any other. The Spanish-speaking people call the plant *Cañutillo*, a word meaning little tube or pipe. Similarly used is the *Encinilla* or Chaparral Tea (*Croton corymbulosus*, Engelm.), a gray-leaved plant of the Euphorbia family found in western Texas and adjacent regions. The flowering tops are the part employed, and an infusion of them is palatable to many. Dr. Havard, in an article on "The Drink Plants of the North American Indians,"[2] stated that in his experience not only Mexicans and Indians enjoyed it, but that the colored United States soldiers of the southwestern frontier preferred it to coffee. The plant contains certain volatile oils but apparently no stimulating principle. *Thelesperma*, a Southwestern

[2] Bulletin Torrey Botanical Club. Vol. XXIII, No. 2.

genus of herbaceous plants of the Composite family, somewhat resembling Coreopsis, with opposite, finely dissected, strong-scented leaves and yellow flowers (sometimes without rays), furnishes a species or two used as substitutes for tea by the Mexican population. *Thelesperma longipes*, Gray, occurring from western Texas to Arizona, is commonly known as Cota, and is said to give a red color to the water in which it is boiled.

Much more appealing to the average taste is a drink that Mexicans sometimes make from the oily kernels of the *jojoba* nut of Southern California and northern Mexico (*Simmondsia Californica*, described previously). Mr. Walter Nordhoff, formerly of Baja California, informs me that the process followed is first to roast them and then treat them in the same way as the Spanish people prepare their chocolate. This, I believe, is to grind the kernels together with the yolk of hard boiled egg, and boil the pasty mass in water with the addition of sugar and milk. When they can afford it a pleasant flavoring is given by steeping a vanilla bean for a moment or two in the hot beverage. This makes a nourishing drink as well as a savory substitute for one's morning chocolate or coffee. A substitute for chocolate among the American population of some sections of the United States is furnished by the reddish-brown, creeping rootstock of the Purple or Water Avens (*Geum rivale,* L.), a perennial herb with coarse, pinnate basal leaves and 5-petaled, purplish, nodding flowers, borne on erect stems a couple of feet high. The plant is frequent in low grounds and swamps throughout much of the northern part of the United States and in Canada, as well as in Europe and Asia. The rootstock is characterized by a clove-like fragrance and a tonic, astringent property, and has been used by country people in decoction as a beverage, with milk and sugar, under the name of Indian Chocolate or Chocolate-root. It is the color, however, rather than the taste that has suggested the common name. Lucinda Haynes Lombard, writing in "The American Botanist" for November, 1918, mentions a curious popular superstition to the effect that friends provided with Avens leaves are able to converse with one another though many miles apart and speaking in whispers!

Readers of literature concerning old time explorations in America will perhaps recall passages in the reports of various writers devoted to accounts of a beverage called Yaupon, Cassena, or the Black Drink, formerly in great vogue among the Indians of the Southern Atlantic States and colonies. One of those ancient chroniclers who did so much to misinform Europe about the New World and its products, speaks of this Black Drink as a veritable elixir that would "wonderfully enliven and invigorate the heart with genuine, easie sweats and transpirations, preserving the mind free and serene, keeping the body brisk and lively, not for an hour or two, but for as many days, without other nourishment or subsistence." (!) William Bartram, to whose account of the Indian uses of Southern plants something over a century ago reference was made in an earlier chapter, speaks of spending a night with an Indian chief in Florida, smoking tobacco and drinking Cassena from conch shells. Bartram does not seem to have liked his Cassena, and in point of fact few white people ever did; but the wide prevalence of its consumption among the Southern Indians, who once drove a brisk inter-tribal trade in the leaves, and the fact that the Cassena plant is nearly related to the famous Paraguyan drink *yerba maté*, have created some latter-day interest in the Black Drink. The plant from which it is made is a species of spineless Holly or Ilex (*I. vomitoria*, Ait.), frequent in low woods from Virginia to Florida and Texas. It is a shrub, or sometimes a modest tree, with small, evergreen leaves which are elliptic in shape and notched around the edge, and in autumn the branches are prettily studded with red berries about the size of peas. An analysis of the dried

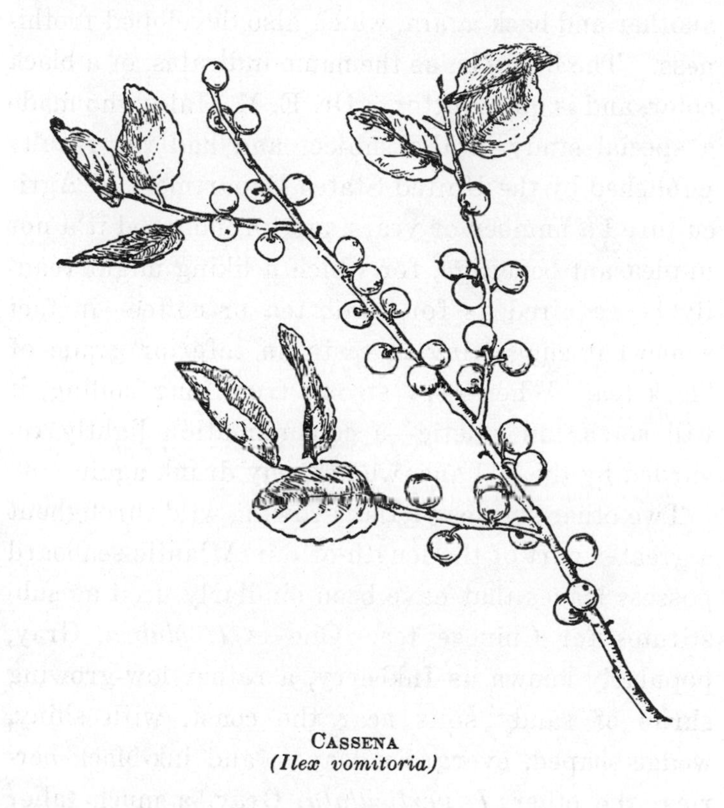

CASSENA
(Ilex vomitoria)

leaves reveals a small percentage (one-quarter of one per cent.) of caffeine, about half the quantity of the same alkaloid that is contained in the leaves of *maté* (*Ilex Paraguayensis*). The leaves were customarily toasted, thoroughly boiled in water, and then cooled by pouring rapidly from one vessel to another and back again, which also developed frothiness. The liquid is, as the name indicates, of a black color, and is quite bitter. Dr. E. M. Hale, who made a special study of the subject and had the results published by the United States Department of Agriculture [3] a number of years ago, pronounced it a not unpleasant beverage, for which a liking might readily be acquired as for *maté*, tea or coffee—in fact somewhat suggesting in taste an inferior grade of black tea. When very strong from long boiling, it will act as an emetic—a consummation lightly regarded by the Indians, who merely drank again.

Two other species of Ilex growing wild throughout a greater part of the length of our Atlantic seaboard possess leaves that have been similarly used as substitutes for Chinese tea. One is *I. glabra*, Gray, popularly known as Inkberry, a rather low-growing shrub of sandy soils near the coast, with shiny, wedge-shaped, evergreen leaves, and ink-black berries; the other, *I. verticillata*, Gray, a much taller shrub, with deciduous foliage, and bright red berries clustered around the stems and persisting in winter.

[3] Bulletin 14, Division of Botany.

The latter species is called in common speech Black Alder or Winter-berry, and frequents swampy ground as far west as the Mississippi.

The spicy, aromatic inner bark and young twigs of the Sweet or Cherry Birch (*Betula lenta*, L.) also deserve mention, as the basis of that old-time domestic brew, birch beer. The characteristic flavor is due to an oil like that distilled from Wintergreen (*Gaultheria procumbens*). This species of birch is a graceful forest tree with leaves and bark suggesting a cherry, and is of frequent occurrence in rich woodlands of the Atlantic seaboard States. The sap is sweet, like the Sugar Maple's, and may be similarly gathered and boiled down into a sugar. The nearly related River Birch (*Betula nigra*, L.), a denizen of low grounds and streamsides throughout much of the eastern United States, particularly southward, is a potential fountain in early spring when the sap is running. At that season, if you stab the trunk with a knife, stick into the cut a splinter to act as a spout, then set a cup beneath to catch the drippings, you will have shortly a draught as clear and cool as spring water, with an added suggestion of sugar. The tree is distinguished by slender, drooping branches, which sleet storms in winter sometimes badly shatter and break. From such untended wounds, hundreds in number, the sap later on will drop pattering to the ground; and I have stepped from bright sunshine on a March day into the shadow of one of these trees and been sprinkled by the descending spray as by a shower of rain.

On re-reading this chapter I see I have overlooked two common wild plants whose possibilities for tea making are worth the camper's knowing. One is that charming little creeping vine with evergreen thyme-like leaves exhaling the fragrance of wintergreen, *Chiogenes hispidula,* T. and G., the Creeping Snowberry, which delights in cool upland bogs of the northern Atlantic seaboard. The tiny white flowers, solitary in the axils of the leaves, are less showy than the white berries which give the plant its name. Readers of Thoreau will recall his brewing his best tea of it in the Maine woods. The other plant is a familiar Pacific Coast fern, *Pellaea ornithopus,* Hook., the Bird's-foot Cliff-brake, found in dry ground nearly throughout California, and easily identified by the division of the fronds into a series of stiff triple-pointed segments strikingly like the three spreading toes of a bird's foot. Tea made by steeping the dried fronds is both tasty and fragrant. Spanish Californians call the fern *calaguala,* and a drink made by boiling root and all is given in tuberculosis and feverish colds.

CHAPTER VIII

VEGETABLE SUBSTITUTES FOR SOAP

To soothe and cleanse, not madden and pollute.
Wordsworth.

AMONG the pleasant pictures of my mental gallery is one of an autumn evening at a Pueblo Indian village in New Mexico, where I chanced to be a few years ago. The sun was near setting, seeking his nightly lodging in the home of his mother, who, according to the ancient Indian idea, lives in the hidden regions of the west; on the house-tops the corn huskers were gathering into baskets the multi-colored ears that represented the day's labor; along the trail from the well some laughing girls were filing, with dripping jars of water on their heads; the village flocks, home from the plain, were crowding bleating into corrals; and from open doors came the steady hum of metates, the fragrance of grinding corn, and the shrill music of the women's mealing songs. Then up the street came pattering a couple of burros loaded with fire-wood and driven by an old Indian man. Immediately three or four women appeared at house doors and called inquiringly "*amole?*" The old man halted his donkeys, lifted from one a sack, out of which he drew several pieces of thick, blackish root, which he distributed impartially among the women, and then proceeded on his way. The root, it transpired, was a sort of vegetable soap and answered to that strange word of the women, *amole.* This, in fact, is the name current throughout our Spanish Southwest for several common wild plants indigenous to that region, and rich enough in saponin to furnish in their roots a natural and satisfactory substitute for commercial soap. Several are species of the familiar Yucca—in particular *Y. baccata, Y. angustifolia* and *Y. glauca.* Americans who prefer their own names for things call them soap-root, when they do not say Spanish bayonet, or Adam's Thread-and-Needle or just Yucca. All three species mentioned have large, thick rootstocks firmly and deeply seated in the earth, so that a pick or crow-bar is needed to uproot them. Before the white traders introduced the sale of commercial soap, *amole* was universally used by Mexicans and Indians for washing purposes, and the practice is not yet obsolete by any means. The rootstock is broken up into convenient sizes and washed free from any adhering dirt and grit. Then, when needed, a piece is mashed with a stone or hammer, dropped into a vessel containing water, cold or warm, and rubbed vigorously up and down until an abundant lather results—and this comes very quickly. After dipping out the fibre and broken fragments, the suds are ready for use. They answer every purpose of soap, and are particularly agreeable in their effect upon the skin, leaving it soft and comfortable. A shampoo of *amole* is, among the long-haired Southwestern Indians, not only a luxury but a prescribed preliminary to ceremonies of the native religious systems. Even whites recognize the efficacy of the root, and an American manufacturer in the Middle West has for years been making a toilet soap with the rootstock of *Yucca baccata* as a basis. It is put upon the market under the name of Amole Soap.

Certain species of Agave, that is, the Century Plant fraternity, are frequent along the Mexican border and contain saponin in greater or less quantity, affording a soap substitute as do the Yuccas. Best known, perhaps, is the species that Spanish-speaking residents call *lechuguilla* (botanically, *Agave lechuguilla*, Torr.). This is distinguished by a cluster of radical, yellowish-green, spine-tipped, fleshy leaves, few in number (rarely over fifteen) and barely a foot long, the flowers borne in a close panicle almost like a spike. The short trunk of the plant is, I believe, the part usually used for soap; but Dr. J. N. Rose, in his "Notes on Useful Plants of Mexico," quotes Havard as authority for the statement that saponin is found in the leaves of this species. The rootstock of a related Texan species (*A. variegata*, Jacobi) is also soapy, and the paper by Dr. Rose just mentioned quotes a statement by a resident of Brownsville, Texas, to the effect that a piece of the rootstock of the latter species as big as a walnut, grated and mixed with a quart of warm water, will clean a whole suit of clothes. The most used Agave-amoles, however, are plants of Mexico, the discussion of which would not be pertinent here.

Of wide occurrence in California is an *amole* of quite a different appearance. It is the bulbous root of a plant of the Lily family, by botanists fearfully and wonderfully called *Chlorogalum pomeridianum*, Kunth. The average American simplifies this into California Soap-plant. Its first appearance is shortly after the winter rains set in, and for several months all that one sees of it is a cluster of stemless, grass-like, crinkly leaves, lolling weakly on the

California Soap-plant
(Chlorogalum pomeridianum)

California Soap-plant
(Chlorogalum pomeridianum)

ground. Late in the spring, a slender flower stalk puts up and at the height of four or five feet breaks into a widely spreading panicle of white, lily-like but small blossoms, that open a few at a time at evening, shine like stars through the night and wither away the next morning. To the economist the most interesting part of the plant is subterranean. This is a bottle-shaped bulb, rather deep set in the ground, and thickly clad in a coat of coarse, brown fibre. When this fibre is stripped off, a moist heart is disclosed an inch or two in diameter and about twice as long. Crush this, rub it up briskly in water, and a lather results as in the case of Yucca and quite as efficacious for cleansing. Indeed, the absence of alkali—an absence that is a characteristic of the *amoles*—makes the suds especially valuable for washing delicate fabrics. Some users of this California amole prefer first to rub the crushed bulb directly upon the material to be washed, just as one would do with a cake of soap, and then manipulate the article in the clear water. The lather is said to be also useful for removing dandruff. However that may be, it unquestionably makes an excellent shampoo and leaves the hair soft and glossy. The bulbs may be used either fresh or after having been kept dry for months. Our knowledge of the cleansing property resident in this bulb is a gift from the California Indian, who, in spite of the popular notion to the contrary, has a taste—though not an extravagant taste—for cleanliness.

Another well-known California soap plant is a species of Pig-weed (*Chenopodium Californicum*, Wats.), abundant throughout much of the State in arroyos and on moist hillsides. It is a stout, weedy-looking herb, with inconspicuous, greenish flowers in slender, terminal spikes, and toothed, triangular leaves turning yellow and dying as the dry season advances. The stout stems, a foot or two high, grow numerously from the crown of a very deep-seated, spindle-shaped root which is at times a foot long and requires industrious digging to lift it from its earthy bed. While fresh it is rather brittle and readily crushed with a hammer, when, if agitated in water, it quickly communicates a soapy frothiness to the liquid, and is cleansing like the other suds noted. The roots may be laid away for use when dry, in which state they are as hard almost as stone, and require to be grated or ground in a handmill before using. The saponaceous property in this root was also discovered first by the Indians.[1]

[1] The roots of the Southern Buckeye or Horsechestnut (*Aesculus Pavia*, L.) are rich in saponin, and Dr. Porcher states that their

A Pacific Coast soap plant (*Chlorogalum pomeridianum*). The bulb, stripped of its fibrous covering, is highly saponaceous. The fiber is useful for making coarse brushes and mattresses.

Tunas, fruit of a Southwestern cactus—Showing how it is opened to secure the meaty pulp. (See page 109.)

The soap plants thus far named must, from the nature of the case, suffer extermination in the fulfilling of their mission, but there are others indigenous to the United States that need not be killed to serve. First among these may be mentioned the genus *Ceanothus*, one species of which—the New Jersey Tea—has already claimed attention in the chapter on Beverage Plants. The genus comprises about thirty-five species, nearly all shrubs or small trees confined to the western United States and northern Mexico. They are particularly abundant on the Pacific Coast, and are popularly known as "wild lilac" and "myrtle" (one or two species as "buck brush"). They are frequently an important element in the chaparral cover of the mountain sides, and in the spring their flowers create beautiful effects in such situations, forming unbroken sheets of white or blue, acres in extent. The fresh blossoms of many species—perhaps of most or even all—are saponaceous, and rubbed in water produce a cleansing lather that is a good substitute for toilet soap. Care must be exercised, however, to pick off any green footstalks that cling to the flowers, as these

suds are preferable to commercial soap for washing and whitening woolens, blankets and dyed cottons, the colors of which are improved by the process.

tend to give the suds a greenish tinge and a weedy smell. This floral soap is not only perfectly cleansing but leaves the skin soft and faintly fragrant. It is a poetic sort of ablution, this bathing with a handful of snowy blossoms plucked from a bush and a little water dipped out of the brook, and revives our faith in the Golden Age, when Nature's friendly outstretched hand was less lightly regarded than nowadays. Similiarly of use are the fresh, green seed-vessels, though these often have a resinous coating that is apt to cause a yellowish stain. Even the foliage of some species is saponaceous.

The cherished Balloon vine of our gardens does not include soapiness among its charms, but it can at least claim cousinship with some of the world's most famous soap plants—namely, certain species of the genus *Sapindus*, trees or shrubs native to the warmer regions of both hemispheres. The name Sapindus means "soap of the Indies," where, as well as in China and Japan, several species have been drawn upon for detergent material from very early times, and are still in favor for washing the hair and delicate goods, such as silk. Within the limits of the United States, three species are indigenous: *Sapindus saponaria*, L., abundant from Brazil to the West Indies, finds a lodgment on the extreme southern tip of Florida, and besides its soapy possibilities possesses seeds, hard and black, that serve for beads and buttons; *S. marginatus*, Willd., an evergreen tree sometimes sixty feet in height, occurs along our southern Atlantic seaboard from the Carolinas to Florida; *S. Drummondii,* H. & A., ranges from Kansas to Louisiana and westward to Arizona, and is known to Americans as Soap-berry or Wild China tree,[2] and to the Spanish-speaking people as *jaboncillo* (little soap). All three species are trees with pinnate leaves (non-deciduous in the first two) and small, white flowers borne in terminal panicles; and all produce fleshy berries about the size of cherries and containing one or two seeds. It is in these berries that the soapy property dwells, and this is readily communicated to water in which the berries are rubbed up. In the case of *S. Drummondii,* the clusters of yellow berries (turning black as they dry) are a conspicuous feature of the bare winter branches, for it is their habit to persist on the trees until spring.

Also of the West is a species of gourd occurring in dry soil from Nebraska to Mexico and westward to the Pacific. In some sections it is known as

2 From its resemblance to the true China tree (*Melia Azedarach*), extensively planted for ornament and shade in the Southern States.

SOAP-BERRY
(Sapindus marginatus)

Missouri Gourd and in California as Mock Orange. Botanically it is *Cucurbita foetidissima*, HBK, and the rank, garlicky odor given off by the crushed leaves makes the specific appellation very apropos. It is a coarse, creeping vine with solitary, showy, yellow flowers and robust, triangular leaves that have a fashion of standing upright in hot weather, like ears; and it spreads so industriously that at the summer's end its tip may be as much as twenty-five feet away from the starting point, which is the crown of a deep-seated, woody, perennial root shaped like a carrot. In the autumn the shriveling leaves reveal numerous, round, yellow gourds, which conspicuously dot the ground and are likely at first glance to deceive one into thinking them spilled oranges—a fact that accounts for one popular name. These gourds are pithy, but such pulp as they contain, as well as in the roots, is saponaceous, and crushed in water both fruit and root yield a cleansing lather. It is, however, apt to leave the skin with a harsh feeling for a few moments, not altogether pleasant. There appears to be saponin in the vine also, since Doctor Edward Palmer has stated that in northern Mexico a Cucurbita, that is undoubtedly this species, has been extensively used by laundresses who mash up the vines with the gourds and add all to their

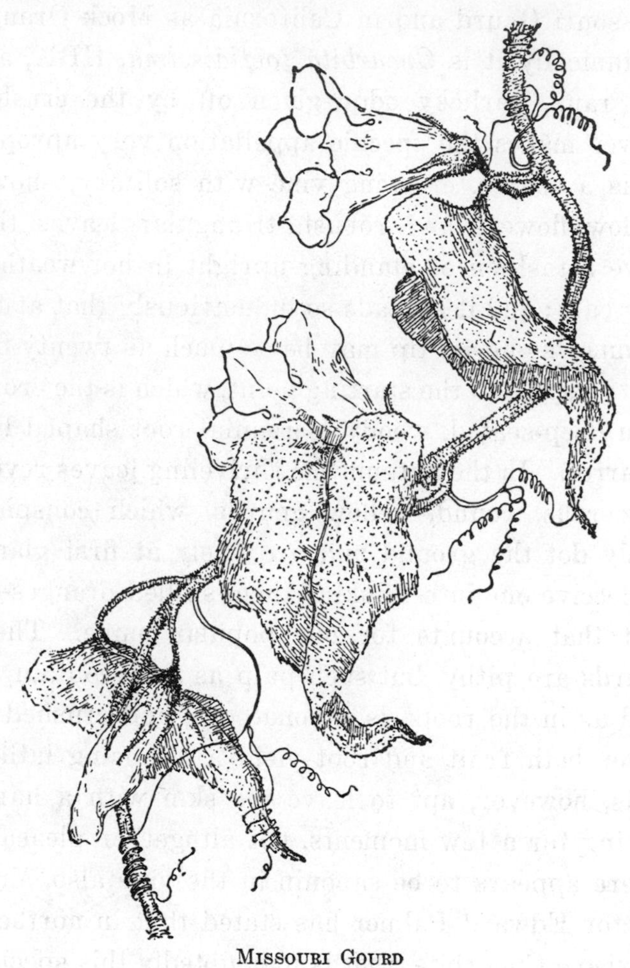

MISSOURI GOURD
(Cucurbita foetidissima)

wash water. To wear under-clothes thus washed, one must be indifferent to the prickles of the rough hairs and broken fibre that are of necessity mingled with the water. Among the Spanish-speaking people of the Southwest, this gourd goes by the name *Calabacilla.* In old plants the root is sometimes six feet long and five or six inches in diameter. This, descending perpendicularly into the earth, enables the plant to reach moisture in arid wastes where shallow-rooted plants would perish. The dried gourds, it may be added, may be very conveniently used as darning-balls.

Probably the most widely known of all our American soap plants—though not all who know the plant are aware that it bears soap in its gift—is an herb of the Pink family that used to have a corner in many old-fashioned gardens under the name of Bouncing Bet (*Saponaria officinalis*, L.). It is a smooth, buxom sort of plant with stems a foot or two tall and noticeably swollen at the joints, oval, ribbed leaves set opposite to each other in two's, and dense clusters of white or pink 5-petaled flowers. It is not a native-born American, but came hither from Europe early in the white immigration and has now become naturalized in many parts of the country near the settlements of men, where it is often so common as to be classed as a weed. The juice of the roots is mucilaginous and soapy, producing a

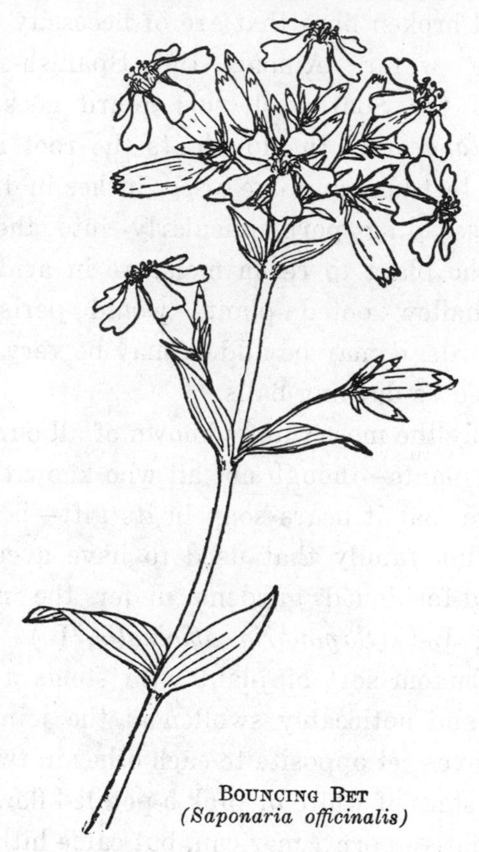

BOUNCING BET
(Saponaria officinalis)

lather when agitated in water, and the peasantry in some parts of Europe use it to-day for soap. By the brothers in European monasteries, centuries ago, its virtue as a capital cleansing agent was well understood, and they employed it for scouring cloth and removing stains. They gave it, in monkish fashion, a Latin name, *herba fullonum*, which in English translation, Fuller's herb, is sometimes still assigned it in books; but in every-day speech the rustic English name, Soapwort, is more usual. In our Southern States a pretty local name that has come to my notice is "My Lady's Wash-bowl." It was in a Saponaria, I believe, that the glucoside saponin—the detergent principle of the soap plants—was first discovered and given its name. That was about a century ago, and since then chemists have identified the same substance existing in varying degrees in several hundred species throughout the world.[3] In most plants, however, the quantity is too small to make a serviceable lather.

[3] N. Kruskal. "Soaps of the Vegetable Kingdom," in "The Pharmaceutical Era," Vol. XXXI, Nos. 13, 14.

CHAPTER IX

SOME MEDICINAL WILDINGS WORTH KNOWING

ROMEO. Your plantain leaf is excellent for that.
BENVOLIO. For what, I pray thee?
ROMEO. For your broken shin.

Romeo and Juliet.

THE subject of medicinal plants is one that I approach with considerable reluctance; because, though the employment of wild herbs as remedies has been a cherished practice with sick humanity whether savage or civilized from the earliest times, there exists still great diversity of opinion about the efficacy of particular simples. One has only to thumb over any ancient herbal or old botanical manual or the succeeding editions of pharmacopœias to notice the decline and fall of one popular medicinal plant after another with the progress of the years, and so to become rather skeptical about the whole subject. Nevertheless, it is a poor chaff-pile that does not hold some kernels of pure grain; and this chapter, without professing to trench upon the province of the chemist who distils and extracts a multitude of medicines from the herbs of the field, will call attention to a few of those plants growing wild whose reputation for the relief of some simple disorders appears well grounded. At any rate they are harmless.

Such medicinal wildings may be classed under two principal heads: those occurring also in Europe or Asia, or naturalized here from the Old World, their uses therefore being part of the white race's traditional knowledge; and those indigenous plants that found place in the medical practice of the Indians, from whom we have got a hint of their value.

In the former class one of the best known is Yarrow or Milfoil (*Achillea Millefolium*, L.), a perennial herb a foot or two high, of the Composite family, with flat-topped clusters of small, usually white-rayed flower-heads, and finely dissected leaves. It is found throughout the United States and much of Canada as well as in Europe and Asia, its medicinal worth, indeed, according to tradition, having first been discovered by Achilles, whence the plant's botanical name. The entire plant above ground may be dried and an infusion of it (a pint of boiling water poured upon a handful) may be administered for a run-down condition or a disordered digestion, the action being that of a mildly stimulating bitter tonic. The familiar Hoar-hound (*Marrubium vulgare*, L.), originally introduced from Europe for a garden herb in the Atlantic States, has long since taken out naturalization papers as an American, and is now found wild across the continent and from Maine to Texas. It is a somewhat bushy perennial of the Mint family, with square, white-woolly stems, grayish, roundish leaves prominently veined and wrinkled, and small, white flowers densely clustered in the leaf axils. The calyx of the flower is provided with ten short teeth hooked at the tips, which catch readily in the coats of passing animals or people's clothing, facilitating the spread of the plant. The dried herb is tonic and a bitter tea made of it is a time-honored household remedy for debility and colds, being expectorant and promotive of perspiration. In large doses it proves laxative.

Apropos of laxatives, an indigenous wild plant that has been popularly esteemed in this regard and whose value was detected because of the herb's relationship to the famous Senna of the Old World, is *Cassia Marylandica*, L., commonly known as Wild or American Senna. The leaves, collected upon the maturing of the seeds, and dried, used to be among the offerings of the Shaker herbalists. An infusion of them may be made in the proportion of about an ounce of the leaves to a pint of boiling water—the dose, two or three fluid ounces of the liquid, repeated

WILD SENNA
(Cassia Marylandica)

if needful. The American plant contains the same general principles as the Old World species but in less proportion, and is correspondingly less active. It is a stout, herbaceous perennial, three to eight

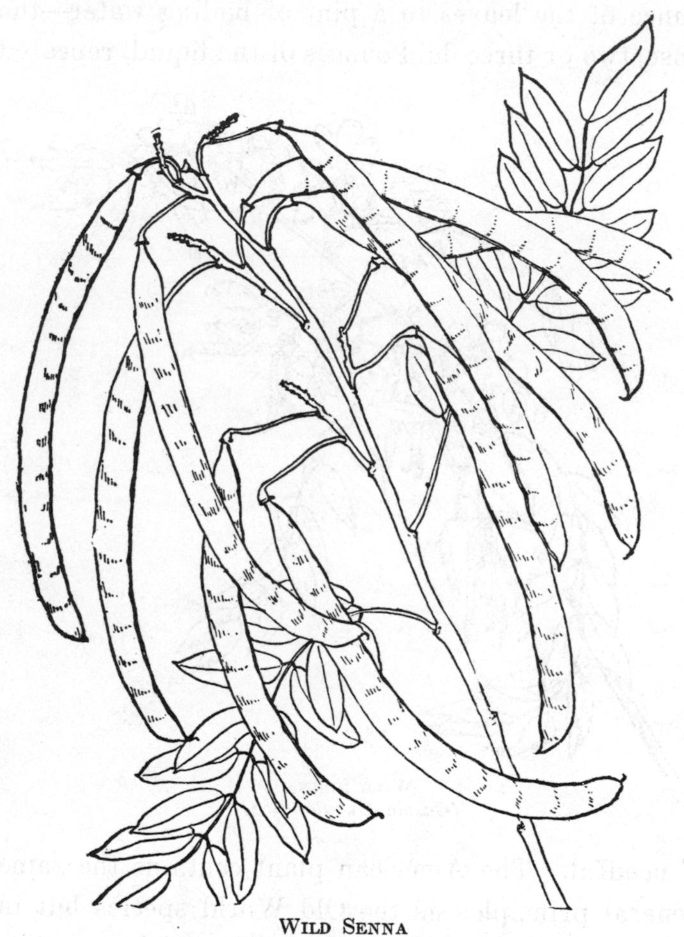

WILD SENNA
(Cassia Marylandica)

feet high, bearing pinnate leaves and showy racemes of yellow flowers in the upper leaf-axils, followed in autumn by long, curved pods or legumes, and occurs in damp ground and swamps from the Mississippi Valley to the Atlantic; and from the Canadian border to the Gulf.

Another plant which, although indigenous, I believe, only to America, is so near akin to a popular tonic herb of Europe that its use may have first been suggested by the resemblance, is Boneset (*Eupatorium perfoliatum*, L.). This is a stout, hairy perennial of the Composite tribe, with rather narrow, pointed, wrinkled leaves opposite in pairs upon the stem and united around it at the base, so as to make each pair present the appearance of one long leaf skewered through the middle; whence another common name for the plant, Thoroughwort. The large clusters of white flower-heads are rayless. The leaves and flowering tops are dried, and a bitter tea is made of them. Taken cold, this is tonic and stimulating in small doses and laxative in large ones. The hot infusion is an old-time remedy for a fresh cold or sore throat, and may be taken during the cold stage of malarial fever. The plant is common in low meadows and damp grounds throughout the eastern United States and Canada.

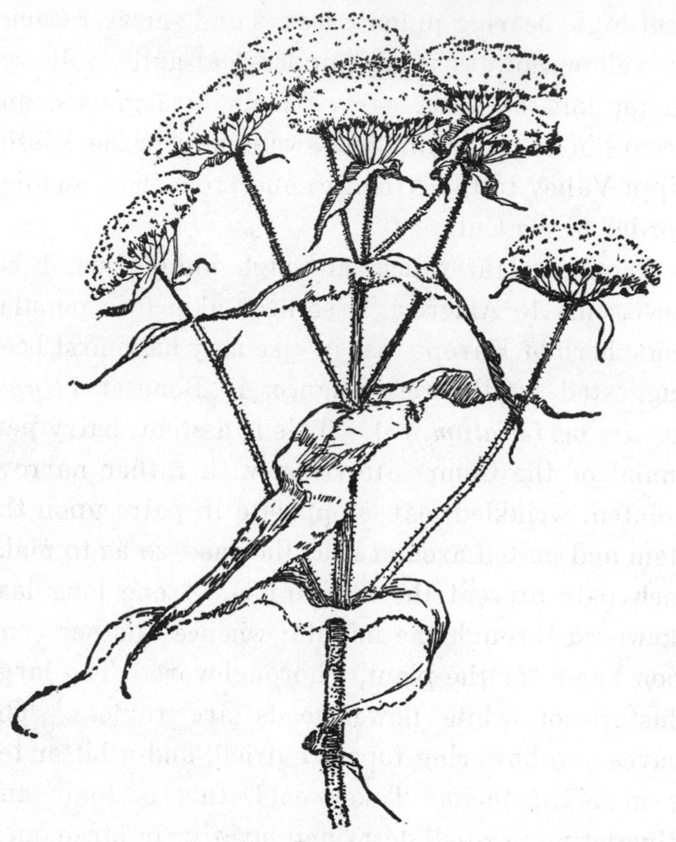

BONESET
(Eupatorium perfoliatum)

And of course every holder to the old traditions is loyal to Wild Cherry bark. This is taken from the familiar Wild Cherry tree (*Prunus serotina*, Ehrh.), growing along streams and fence-rows and in woods from eastern Canada to Texas. It is from forty to eighty feet high and identifiable by its shiny green leaves (too often a prey to caterpillars) and

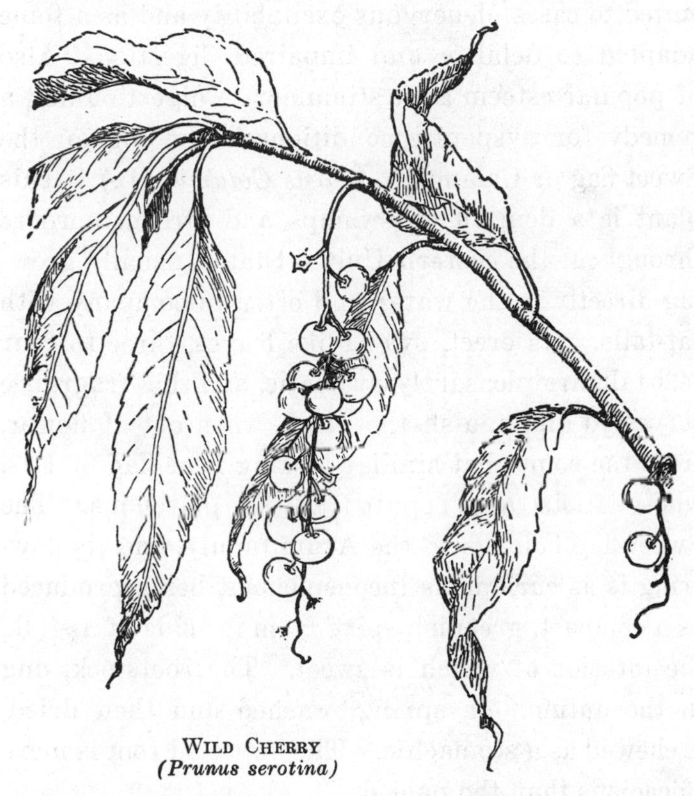

WILD CHERRY
(Prunus serotina)

its close racemes of small white flowers succeeded by small, black, juicy, flattened fruit with a bitter but vinous flavor. An infusion of the dried bark (gathered preferably in the autumn) in cold water, in the proportion of one-half ounce of bark to a pint of water, enjoys a reputation both as a mild sedative suited to cases of nervous excitability and as a tonic adapted to debility and impaired digestion. Also of popular esteem as a stimulant to digestion and a remedy for dyspeptic conditions is the root of the Sweet-flag or Calamus (*Acorus Calamus*, L.). This plant is a denizen of swamps and stream borders throughout the eastern United States, usually growing directly in the water and often in company with cat-tails. Its erect, sword-like leaves, three to four feet tall, are pleasantly aromatic, and this fragrance serves to distinguish the plant, when out of flower, from the somewhat similar-looking Blue-flag or Iris, whose roots are reputed to be poisonous. The Sweet-flag belongs to the Arum family, and its flowering is as curious as inconspicuous, being produced as a compact, greenish spike from the side of a stalk, the interior of which is sweet. The rootstock, dug in the autumn or spring, washed and then dried, is chewed as a stomachic. The unpeeled root is more efficacious than the peeled.

It was the popularity of the Old World Pennyroyal doubtless that first caused attention to be directed to a little minty annual common in dry soil and old fields pretty much throughout the United States east of the Mississippi and called American Pennyroyal (*Hedeoma pulegioides*, Pers.). It is pungently aromatic, from a few inches to a foot tall, with small, opposite leaves narrowing to the base and tiny, bluish flowers clustered in the upper leaf-axils. The plant contains a volatile oil, and a hot infusion of the dried leaves and flowering tops is an old-fashioned remedy for flatulent colic, sick stomach and bowel complaints. Then there is the nearly related Dittany (*Cunila Mariana,* L), growing on dry woodland hills from New York to Florida, a perennial plant of about the height of the American Pennyroyal, but with larger leaves, rounded at the base and conspicuously clear-dotted. The herb is gently stimulant, and a tea made of it is a pleasant

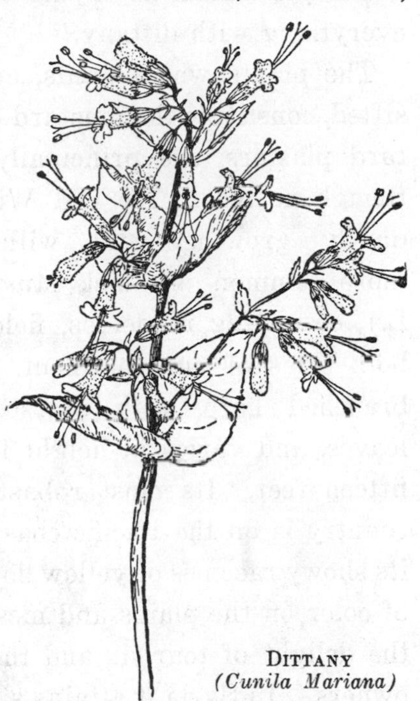
DITTANY
(Cunila Mariana)

and refreshing beverage that is sudorific and has a respectable place among the rural remedies for febrile conditions. Dr. Porcher quoted an old-time South Carolinian as saying that "everybody cured everything with dittany."

The plants whose seeds, crushed to a flour and sifted, constitute the mustard of commerce and mustard plasters, are principally two, both of which, though native to the Old World, are found abundantly growing wild within our limits. The more common is Black Mustard (*Brassica nigra*, L.), occupying roadsides, fields and waste land on both sides of our continent. It is a stout, much-branched herb, with coarse, deeply lobed basal leaves, and varies in height from two to twelve or fifteen feet. Its most robust development in this country is on the Pacific coast, where in the spring its showy racemes of yellow flowers make solid sheets of color on the plains and mesas, acre upon acre, to the delight of tourists and the disgust of the land-owners. In Syria it attains similar proportions and is believed to be the mustard of the gospel parable. The other Mustard plant is the closely related *Brassica alba*, (L.) Boiss., popularly known as White Mustard. It is rarely over two feet high, and is distinguished from its black cousin by hairiness of stem and seed pod, the latter curiously ending in a broad, flat, sword-shaped beak.

Among a considerable portion of our population the Indians have enjoyed from very early times a reputation for special knowledge in the remedial properties of wild plants; but doubtless they have been credited much in excess of their deserts. Nevertheless, there are some of the aboriginal remedies worthy of all respect. Prominent among them are two or three plants of the Pacific Coast. One of these seems first to have been brought to light through the contact of the Franciscan missionaries of the eighteenth century with the Indians of Southern California, and is still quite generally known by its Spanish name, Cáscara sagrada, that is "sacred bark." It is a shrub or small tree of the genus *Rhamnus*, with somewhat elliptic, prominently veined leaves, abundant clusters of tiny yellowish flowers in spring succeeded in the autumn by a conspicuous crop of inedible berries turning yellowish-crimson and finally black. The plant is considered by some botanists as of one variable species (*Rhamnus Californica*, Esch.), and by others as of two—the name *R. Purshiana*, DC., being applied to the arboreal form, which is common through the northern coast regions as far as British Columbia and east-

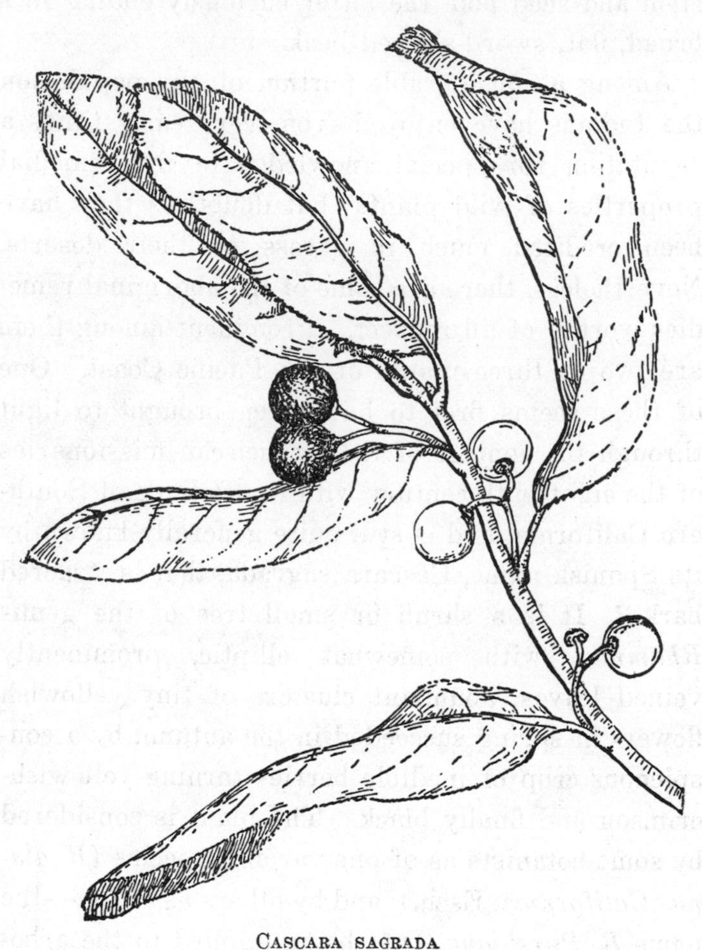

CASCARA SAGRADA
(Rhamnus California)

ward to the Rockies, attaining a height at times of thirty feet or so, with a trunk a foot in diameter. In that region it goes by a number of names as Chittem-wood, Wahoo and Bitter-bark. Other local names are Pigeon-berry and Wild Coffee—the latter because of some superficial resemblance of the seeds to coffee beans. The shrubby form, common in Southern California and the Great Basin region, is from a few to a dozen feet high, forming usually a dense clump touching the ground.

The medicinal value of the Cáscara sagrada is in the bark, which is regarded as one of the safest and best laxatives in the world, especially valuable in cases of chronic constipation. It acts, at the same time, as a tonic and tends to improve the appetite. For the best results the bark should be collected in the autumn or early spring and at least a year before being used. A small piece of the bark put into a glass of cold water and allowed to soak over night makes a useful tonic, drunk first thing in the morning. For a laxative, hot water should be poured upon the bark in the proportion of a teacupful to a level teaspoonful of the finely broken bark, set away to cool, and drunk just before bed-time. Country people have told me that the fresh bark boiled several hours is equally efficacious. The gathering of

Cáscara sagrada for the medical trade is an important minor industry in the Pacific Northwest, the bark of the *Purshiana* or arboreal form being the kind preferred. There is a considerable European demand for it, as well as from American chemists.

Another of the famous Pacific Coast remedies is Yerba Santa, whose Spanish name (meaning "holy herb") also betrays its connection with the California Mission days, when the Padres not only instructed Indians but now and then learned something from them. An American common name for the plant—Consumptive's Weed[1]—indicates one of its popular uses. It has, in fact, been esteemed for generations in California as an expectorant, a blood purifier, and a tonic—a standby in all bronchial and respiratory troubles. Botanically it is *Eriodictyon glutinosum*, Benth., and is a shrubby plant, three to seven feet high, with dark green, resinous leaves (shaped somewhat like those of the peach) glutinous and shining on the upper side and whitish underneath, the flowers tubular, clustered and usually purple but sometimes white. It is abundant on dry hillsides and among the chaparral throughout much of California and southward into Mexico. A bitter

[1] Others are Mountain Balm, Gum Leaves, Bear's-weed and Wild Peach.

YERBA SANTA
(Eriodictyon glutinosum)

tea is made of the dried leaves and taken freely; or it may be prepared by boiling with sugar, if it is desired to disguise the bitterness. The pounded leaves have also been used as a poultice, bound upon sores.

The civilized drug Grindelia is derived from certain species of a botanic genus of that name belonging to the Sunflower family and occurring rather abundantly on the plains and dry hillsides west of the Mississippi. They are coarse, sticky plants, characterized by white, gummy exudations upon the buds and flower heads (these latter are conspicuously yellow-rayed) and are popularly called, on that account, Gum-plants. The California Indians are credited with being the pioneers in discovering the remedial secret of these plants, the species most used by them being apparently *Grindelia robusta*, Nutt. A decoction of the leaves and flowering tops collected during the early period of bloom is a mild stomachic, and is taken to purify the blood, as well as to relieve throat and lung troubles.

The Indian is also to be thanked for our knowledge of Yerba Mansa (or more correctly, Yerba del Manso, "the herb of the tamed Indian"), common in wet, alkaline soil throughout much of the South-

YERBA MANSA
(*Anemopsis Californica*)

west—a low-growing perennial, carpeting the ground with its dock-like leaves and starred in spring with conical spikes of small, greenish florets, subtended by showy involucres of white bracts. It is the botanists' *Anemopsis Californica,* H. & A. The peppery, aromatic root is astringent, and is chewed raw, after drying, for affections of the mucous membrane, and also made into a tea for purifying the blood. It is one of the most popular of remedies among the Mexican population, who employ it also to relieve coughs and indigestion or pretty much anything. As an external remedy for cuts, bruises and sores on man or beast, either the tea or a poultice of the wilted leaves is employed.

For external use in such cases, two other western plants are valuable, particularly for the healing of that bane of the horseman, the saddle gall. One is an ill-smelling shrub of the Southwestern desert region variously called Creosote-bush, Greasewood (one of many Greasewoods, by the way) and, by its Spanish names, Gobernadora and Hediondilla. Botanically, it is *Larrea Mexicana,* Moric., or, according to other nomenclaturists, *Covillea tridentata,* (DC.) Vail. It is distinguished by curious little evergreen leaves each consisting of two pointed, sticky leaflets, yellow 5-petaled flowers, the petals set edgewise to the light, and round silky seed-vessels like fluffy white pellets. The branches are banded at intervals in black. It grows in the aridest of soils, from Southern California eastward

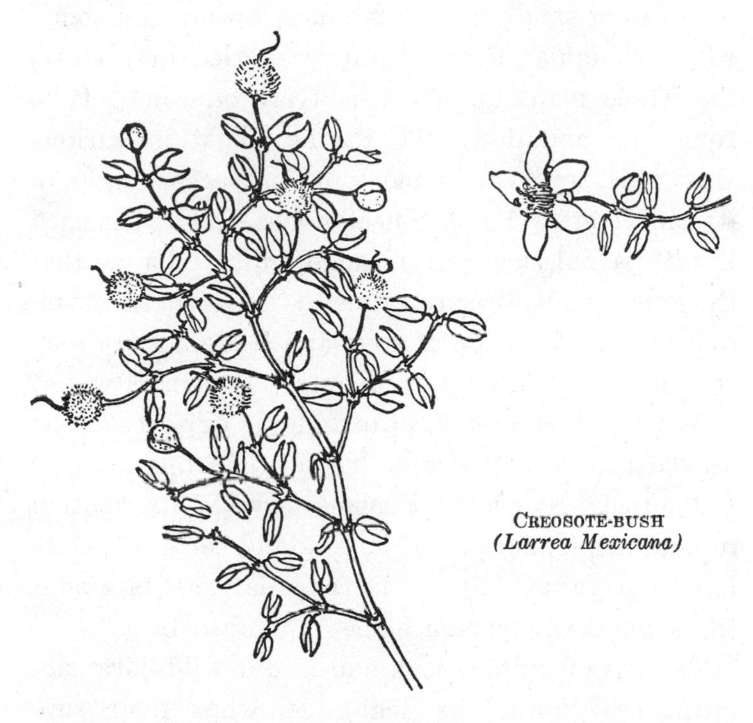

CREOSOTE-BUSH
(Larrea Mexicana)

across Arizona and southward into Mexico. An antiseptic lotion may be made by steeping the twigs and leaves in boiling hot water, effective in the treatment of sores and wounds both of men and animals.[2] The other plant referred to is *Stachys Californica,* Benth., called Mastransia by the Mexicans, with whom it is a standard remedy. It is a hairy herb of the Mint tribe, a foot or two high, with rather small, purple, 2-lipped flowers and somewhat triangular leaves rather wrinkled in texture, the whole plant quite distinctively odorous. It is found up and down the Pacific Coast in various situations, and varies more or less accordingly in its characters. Mr. J. Smeaton Chase, who has used it with signal success for saddle galls, tells me that the green plant, freshly gathered, is customarily employed. An infusion of stem and leaves is made by soaking them for a few minutes in boiling water. This is applied as a wash to wounds or sores. The soaked leaves may also be bound upon the parts as a poultice. *Stachys* is a genus of wide distribution in both hemispheres, and in England certain species long ago gained repute as remedial agents, under the suggestive common name Woundwort.

Patrons of quinine may find in our wild flora substitutes by no means negligible, when their supply of cinchona gives out. The most important are

[2] Mr. J. S. Chase, in his recent book "California Desert Trails," states that a half inch or so of the stem of the Creosote-bush, peeled and held in the mouth like a pebble, is an Indian device for staving off thirst on desert journeys when water is scarce.

Flowering Dogwood (*Cornus florida, L.*) The bark is used in making a medicine similar to quinine, and that of the root produces a red dye used by the Indians. (See page 225.)

(***Courtesy of the New York Botanical Gardens.***)

certain shrubs or small trees of the Dogwood family, which has representatives on both sides of the continent. One of these is the well-known Flowering Dogwood (*Cornus florida,* L.), which beautifies spring woodlands with its showy white floral involucres from Canada to Florida and Texas. The bark is tonic, mildly stimulant and anti-intermittent, and many physicians have recognized its worth as a remedy in intermittent fevers, inferior only to Peruvian bark. A decoction is made of the dried bark of either the tree itself or the root, the latter being the stronger. (The fresh bark is said to be cathartic.) On the Pacific Coast from British Columbia to Southern California a kindred species is the Western Dogwood (*Cornus Nuttallii,* Aud.), which resembles in general appearance its eastern cousin. The bark is similarly useful. Townsend, in his journal of the Wyeth expedition to the Pacific Coast in the early days, tells of his curing two Oregon Indian children of fever-and-ague with this Dogwood, his supply of quinine being exhausted. He boiled the fresh bark in water and administered about a scruple a day. In three days his little patients were well. As he worked over the decoction, the Indians crowded about him curiously; and "I took pains," he writes, "to explain the whole matter to them, in order that they might at a future time be enabled to make use of a valuable medicine which grows abundantly everywhere throughout the country."

Closely related to the Dogwoods is a genus of shrubs called by botanists *Garrya.* Several species are indigenous to our Far West. They are evergreen with inconspicuous flowers, which are of two sexes borne on separate individuals in drooping, tassel-like clusters or catkins. *Garrya elliptica,* Dougl., is a common shrub of the California chaparral, that has been considered ornamental enough to be introduced into gardens both in this country and abroad under the name "Silk-tassel bush." Bark, leaves and fruit are exceedingly bitter. The inherent principle seems to be the same as in the Dogwoods, and a decoction of bark or leaves has been similarly used for the relief of intermittent fevers. The shrub is known locally as Quinine-bush and Fever-bush.[3]

[3] A multitude of wild plants have at various times and in all parts of our country had a place in popular favor as remedies more or less efficacious for the bite of venomous serpents. They are usually called, in common speech, Rattlesnake-weed, Rattlesnake-root, Rattlesnake-master, or among the Spanish-speaking people of the Southwest, *Yerba de Víbora* or *Golondrina.* Their real value, however, is so questionable that it seems hardly worth while to devote space here to their description.

Among Spanish Californians an herb of the Pacific Coast believed useful in fevers is Canchalagua, or as the Americans call it Wild Quinine (*Erythraea venusta,* Gray). It is of the Gentian family, whose characteristic bitterness it possesses; and is one of the most charming of western spring flowers, common on dry hillsides throughout much of California—the bright pink blossoms with a yellow eye borne in terminal clusters upon plants a few inches to two feet high, with lance-shaped leaves in opposite pairs. Of the same family and somewhat similar in appearance but with leaves clasping a quadrangular stem is the American Centaury (*Sabbatia angularis,* Pursh.), common on the Atlantic side of the continent from Canada to Florida. The dried herb is intensely bitter, and is popular among old-fashioned folk for its tonic properties.

One of the most interesting plants of the Pacific Coast is a beautiful evergreen forest tree, known variously as California Bay, California Laurel, Pepperwood and Oregon Myrtle (*Umbellularia Californica* [H. & A.] Nutt.). It is a member of the Laurel family (to which the Sassafras, the Old World Bay and the Camphor-tree belong) and is characterized by a strong, pungent odor given off from the crushed leaves, somewhat suggesting bay

CANCHALAGUA
(Erythraea venusta)

rum. This peculiar aromatic quality of the leaf is due to a resident volatile oil and has the effect of causing headache if inhaled too freely. On the homeopathic principle Northern California Indians, according to Chesnut, sometimes place a bit of leaf in the nostril to cure headache. A decoction of the foliage has been used as a disinfectant and insecticide, and hot baths with bay leaves thrown in, followed by a rubdown, are an Indian remedy for rheumatism.

Goldenrod, so abundant in the United States as to be almost the national flower, derives its name, however, from the European species *Solidago virgaurea* (occurring also but sparingly along the southern Canadian border). The plant is tonic and astringent with a special reputation for healing wounds, and Linnaeus preserved this idea in his name for the genus, *solidago* meaning "I make whole." Spanish Californians make a lotion of the boiled leaves and stems of *Solidago californica,* Nutt., for sores and cuts on man or beast, finishing off with a sprinkling of the powdered leaves. Their name for the species is *oreja de liebre,* jack-rabbit's ear, from a fancied resemblance in the leaf. The tea from *S. odora,* mentioned on page 147, is sometimes prescribed for flatulence and colic.

CHAPTER X

MISCELLANEOUS USES OF WILD PLANTS

> O mickle is the powerful grace that lies
> In plants, herbs, stones, and their true qualities;
> For nought so vile that on the earth doth live
> But to the earth some special good doth give.
>
> *Romeo and Juliet.*

IN the days before game laws came into being within the limits of the United States, several wild plants were employed for catching fish. I do not mean that they were used as bait, but in a very different way, long practised by the Indians. The plants in question contain in their juices narcotic poisons, which, stirred into the water of ponds, deep pools or running streams temporarily dammed, containing fish, stupefy the latter without killing them, and cause them to float inert to the surface, where they may be easily gathered into baskets. No ill effects appear to result from eating fish so poisoned, and in old times in California there was ample chance to test the matter, as both white men and red were prone to satisfy their appetite for fish in this manner. Such pot-hunting has now, however, for many years been forbidden by law. In California the bulbs of the Soap-plant (*Chlorogalum pomeridianum,* already described) were mostly used, being first crushed in quantity, thrown into the water, and mixed with it. Next to these in popularity were the macerated stems and leaves of the Turkey Mullein (*Croton setigerus,* Hook.), the Spanish-Californians' *Yerba del pescado*—that is, "fish-weed." This plant is a rather low-spreading, bristly-hairy, grayish herb, with little greenish blossoms that are scarcely noticeable. It appears in the fields and plains of midsummer and remains through the autumn. Hunters of wild doves know it well, as these birds are very fond of the seeds and collect in numbers to feed where the "mullein" grows—to their undoing. Employed in the same way on the Atlantic seaboard were the seeds of the Southern or Red Buckeye (*Aesculus Pavia,* L.), a tree that occurs from Virginia to Florida and westward to the Mississippi Valley. According to Porcher, the fresh kernels were customarily macerated in water, mixed with wheat-flour to form a stiff paste, and thrown into pools of standing water. The dazed fish would float up to the top and had then only to be picked up. If placed in fresh water, they would soon revive.

When they wanted to, Indians knew quite well where to go for material for fishing lines and nets—their knowledge of wild plants packed with useful fiber being rather extensive. One of the most widely distributed of these native fiber plants is the so-called Indian hemp (*Apocynum cannabinum*, L.), an herbaceous perennial with a smooth, milky-juiced, woody stem two to four feet high, and inconspicuous, greenish-white flowers producing very slender seed-pods about four inches long. It is found in thickets and dampish ground from Canada to Mexico and from the Atlantic to the Pacific. The usual preliminary preparation—as in the case of all the wild fiber-plants, I believe—was to rot the stems by soaking them in water. After that the outer

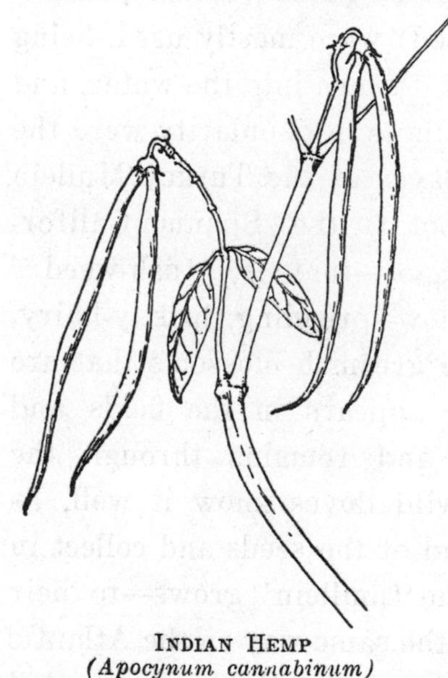

INDIAN HEMP
(Apocynum cannabinum)

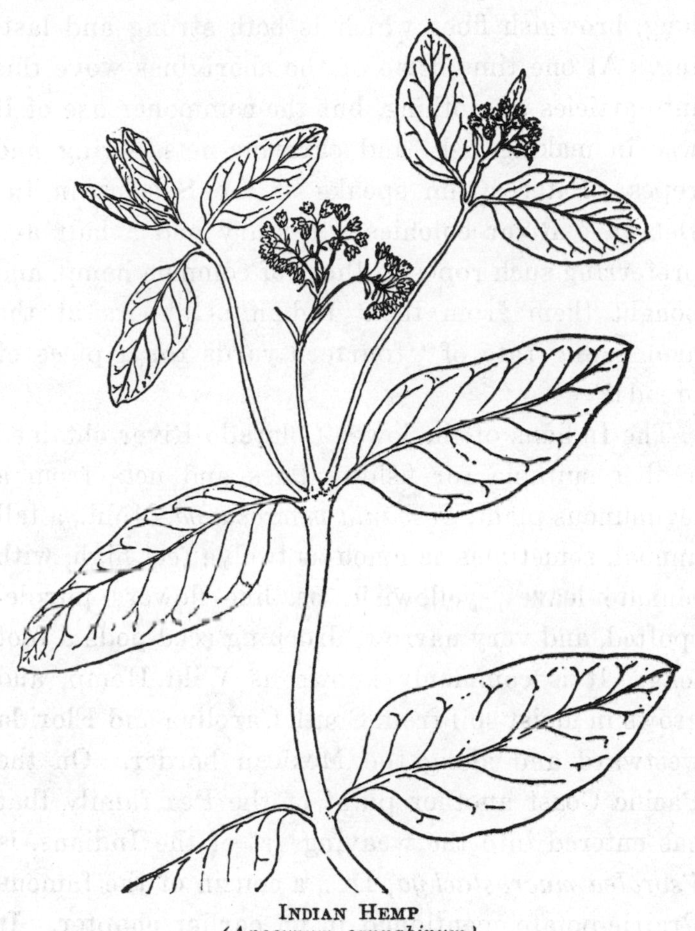

INDIAN HEMP
(Apocynum cannabinum)

bark readily separates and leaves exposed a soft, long, brownish fiber which is both strong and lasting. At one time some of the aborigines wove this into articles of clothing, but the commoner use of it was in making fish- and carrying-nets, string and ropes. Peter Kalm speaks of the Swedes in the Delaware River colonies a century and a half ago preferring such ropes to those of common hemp, and bought them from their Indian neighbors at the astonishing rate of "fourteen yards for a piece of bread!"

The Indians of the lower Colorado River obtained a fiber suitable for fishing lines and nets from a leguminous plant, *Sesbania macrocarpa,* Muhl., a tall annual, sometimes as much as twelve feet high, with pinnate leaves, yellowish, pea-like flowers purple-spotted, and very narrow, drooping seed-pods a foot long. It is commonly known as Wild Hemp, and grows in moist soil from South Carolina and Florida westward and along the Mexican border. On the Pacific Coast another plant of the Pea family that has entered into the weaving art of the Indians, is *Psoralea macrostachya,* DC., a cousin of the famous Prairie-potato mentioned in an earlier chapter. It is a stout, heavy-scented perennial, three to twelve feet high, with leaves consisting of three leaflets, and bearing in summer silky spikes of small, purplish flowers. Its favorite habitat is the borders of streams. Besides the inner bark, which is an excellent material for making coarse thread, the large root contains a valuable fiber. This the California Indians used to secure by pounding out the root. A pleasing feature of the fiber, whether of the root or the stem, is an aromatic perfume, which persists for months.[1] Various species of Nettle, too, soaked in water, yield a fiber for cord making, as the Indians long since discovered. The Nettle, indeed, has been a primitive source of thread in both hemispheres; and Prior, in his "Popular Names of British Plants," quotes an old writer as saying, "Scotch cloth is only the housewifery of the nettle."

Another fairly good fiber, utilizable for twine and rope, has been secured from several species of *Asclepias,* the familiar Milkweeds. Among these may be mentioned especially the Swamp Milkweed (*Asclepias incarnata,* L.), with smooth stem and foliage, and red or rose-purple flowers. It is a frequent denizen of swampy land throughout the eastern half of the country from Canada to the Gulf. In the same class is a well-known woolly Milkweed

[1] Chesnut, "Plants Used by the Indians of Mendocino Co., California."

of the Pacific Coast (*A. eriocarpa,* Benth.), characterized by cream-colored flowers and foliage clothed with a hoary hairiness. The commonest Milkweed of eastern fields and waste places, *A. Syriaca,* L., yields a fiber that has been used to some extent in paper making, and for weaving into muslins. In fact, the white man's interest in all our wild fibers has been largely directed in latter times to their adaptability to adulterating and cheapening fabrics.[2]

The most important of all our native fiber plants are the Yuccas and Agaves. It is from Mexican species of the latter genus—and possibly of both genera—that the valuable Sisal-hemp, imported from Mexico, is made, with which our United States species have never successfully competed. Fiber from the Yucca (probably *Y. baccata,* Torr.) was in extensive use by the prehistoric people who built the cliff dwellings of the Southwest, as is proved by sandals, rope and cloth found in these remarkable ruins. According to the Zuñi tradition it was from Yucca fibers that men made the first clothing for

[2] For many interesting details touching the general subject of wild fibers, reference is made to Reports 5 and 6, Office of Fiber Investigation, U. S. Dept. of Agriculture, entitled respectively "Leaf Fiber of the United States," and "Uncultivated Bast Fibers of the United States," by C. H. Dodge.

themselves when they emerged from the underworld (their first home) into this world of light. Though the spread of white education among our aborigines has caused this ancient textile art to become almost a lost one, it is not entirely so. Here and there an old Indian is still run across who holds to the traditions of the elders and works the ancient works. One such not long ago, living on the California desert, made me from the fiber of the Mescal plant (*Agave deserti*) a pair of sandals of immemorial pattern, the spongy sole an inch thick turned up at the heel, and with an elaborate arrangement of cords to keep the foot in place.

Both Agave and Yucca are treated in the same manner to separate the fiber. After soaking the leaves in water to soften them, they are pounded and repeatedly rinsed until the pulpy part is disposed of. The fibers are then combed out, twisted into strands, and woven as desired. According to Dr. Palmer, the old-time Southern California weavers were famous for their Yucca fiber ropes, nets, hairbrushes and saddle blankets. In the last a padding of softer fiber obtained from the quiote (*Yucca Whipplei*) was employed to relieve the harshness of the *Yucca baccata* fiber.[3] The tough

[3] The American Naturalist, Sept., 1878.

epidermis of Yucca leaves, split into narrow strips, makes a coarse basket material, serviceable moreover as a cord substitute for tying and jacketing articles to be hung up, as hams and watermelons. In the East the same may be done with the strong, fibrous bark of the Moose-wood or Leather-wood (*Dirca palustris,* L.), the *bois de plomb* of the French-Canadians. It is a deciduous shrub, two to six feet high, much branched and characterized by a tough bark, suggesting leather in its pliability, the pale greenish flowers preceding the leaves in small terminal fascicles in early spring. Damp woodlands are its favorite home, from Canada to the Gulf and eastward from the Mississippi to the Atlantic.

A good string may also be made by twisting the fiber obtained from the common Reed-grass (*Phragmites communis,* Trin.),—the Carrizo of the Southwest,—whose tall, straight canes crowned with silky, plume-like floral panicles, form a conspicuous feature in swamps and damp places throughout the United States and Canada. At a distance they present the general appearance of Broom-corn. A peculiarity of this reed that excited the curiosity of observant explorers half a century or so ago, was utilized by some of the Indian tribes to minister to their taste for sugar. Owing to the attacks of a certain insect, which punctures the leafage, a pasty exudation is often to be found in abundance upon the plants. This, upon hardening into a gum, may be collected, and has a sweet, licorice-like taste. Palmer records a former practice of the Indians to cut the canes when the gum was sufficiently hardened, lay them in bundles upon blankets, and shake off the sweet particles. The sugar thus obtained was usually consumed by stirring it in water, making thus a sweet and nutritious drink. Coville speaks of a somewhat different practice with the same plant by the Panamint Indians of the Mojave Desert, who would dry the entire reed, grind it and sift out the flour. This, which would be moist and sticky from the inherent sugar, would then be set near a fire until it would swell and brown, when it would be eaten like taffy.[4]

Another primitive sort of sugar harvest may be reaped in a small way from the common Milkweed (*Asclepias Syriaca*). Kalm, among others, has noted this. The process as observed by him was to gather the flowers in the morning while the dew was on them. The dew, expressed and boiled, yielded a palatable brown sugar. Such a dainty sort of manufacture seems fitting enough in fairy

[4] The American Anthropologist, Oct., 1892.

economics; but it is hard to believe it to have been of much practical value among the rough pioneers from whom the old Swedish traveler learned of it. The Sugar Pine (*Pinus Lambertiana,* Dougl.), that noblest of Pacific Coast pines, owes its common name to a sugary exudation from the heart-wood when the tree has been cut into with an ax or been damaged by fire. The bleeding sap forms irregular lumps and nuggets, white when fresh and unstained, but more often found brown from exposure and contact with fire. John Muir thought this sugar the best of sweets. As to that, each must be his own judge; but it certainly has an appeal to many. Moderation should be exercised in its consumption, as it has a decided laxative tendency. Of all "wild sugars," however, the sap of the Sugar Maple, the source of commercial maple sugar, is without a peer. It is too well known to call for more than mention here.

Our wild plants that have been experimented upon for dyes by the color-loving Indians are very numerous. The subject is too technical for me to say just what value these various vegetable dyes may have in the arts of civilization, but I may refer briefly to a few.

Imprimis, there is that familiar hedge-plant, the

Osage Orange (*Maclura aurantiaca,* Nutt.). Its native home is in the rich bottom-lands of a comparatively narrow strip of territory extending from eastern Kansas and Missouri through Arkansas to Texas, attaining in all that region arboreal proportions. It is distinguished by its curious, yellowish-green, rough-skinned, milky, but inedible fruits, somewhat resembling half-ripe oranges. The large roots and the heartwood of the tree are bright orange in color, and from the former has been extracted a yellow dyestuff, which has been pronounced comparable in excellence to fustic, the product of an allied tree of the tropics. The elastic, satiny wood was a favorite material for bows among the Indians,[5] and the tree came to be known accordingly by the French-Louisianians as *Bois d'arc.* A curious use of the milky juice of the "oranges" is recorded by Dr. James of the Long expedition, the members of which resorted to smearing themselves with it as a protection from the torment of wood-ticks.

From Kentucky to North Carolina, the beautiful Kentucky Yellow-wood (*Cladastris tinctoria,* Raf.) is indigenous, a smooth-barked tree with pinnate

[5] "The price of a bow made from this wood, at the Aricaras', is a horse and blanket." John Bradbury's "Travels in the Interior of America." 1809–11. But the Aricaras lived a thousand miles from where the Osage Orange grows.

leaves and showy panicles of fragrant, white, pea-like blossoms, pendent in June from the branch ends. It, too, has yellow wood, as the common name implies, and from it a clear saffron dye may be had. Better known is the Quercitron or Dyer's Oak (Bartram's *Quercus tinctoria*), which has played a part in international commerce. The inner bark, which is orange-colored, yields a fine yellow dye, and was once an important article of export to Europe, where it was employed in the printing of calicos. The tree is indigenous in poor soil throughout a large part of the eastern United States, and by some botanists is regarded as but a variety of the Scarlet Oak (*Quercus coccinea,* Wang.), whose foliage is a fiery contributor to the autumn coloring of our forests.

Nature's fondness for yellow is manifested in her gift of many dyes of this cheerful color, utilized by her red children. The common Wild Sunflower (*Helianthus annuus,* L.) and the flower heads of the rank-smelling Rabbit-brush (*Chrysothamnus nauseosus* [Pursh.] Britt.)—this latter one the commonest shrubs of the Far Western plains and deserts, with rayless flat-topped clusters of yellow flowers and with linear leaves—have long yielded a yellow stain to the Indians, who transmute the gold of the blossoms into liquidity by the process of boiling. Another mine of color is Shrub-yellow-root (*Xanthorrhiza apiifolia,* L.Her.), a low, shrubby plant of the Buttercup family, with pinnate leaves clustered at the top of a short stem, and small, brownish-yellow flowers in drooping, slender racemes appearing in April or May, in woods and on shady banks of mountain streams from New York to Florida. The bark and roots are richly yellow, and from the latter the dye was customarily extracted. The bark and roots, too, of some of the Barberries (notably the western *Berberis Fremontii,* Torr.) yield a yellow dye, of which the Navajos used to be fond as a color for their buckskins. Equally in aboriginal favor as a source of yellow was the nearly related Golden Seal (*Hydrastis Canadensis,* L.), the thick, orange-colored rootstock being used. It occurs in rich woods from the Canadian border to Arkansas and Georgia—a low herb, with a hairy stem two-leaved near the summit which bears a single, greenish-white flower. It is sometimes called Yellow Puccoon.[6]

Puccoon is a word of Indian origin, and has been applied to other plants as well. One of these, the Red Puccoon, is more commonly known as Bloodroot (*Sanguinaria Canadensis,* L.), whose hand-

[6] The root is also the source of the official drug Golden seal, and its collection on this account has caused the plant to become exterminated in many localities where it was once common.

some, white flowers are among the best beloved of the woodland posies of spring, from Manitoba to Florida. The whole plant is charged with a bitter juice of a reddish-orange color, and that of the rootstock was used by the Indians to produce a bright red coloring matter with which they painted their bodies, and also colored articles of native manufacture, particularly baskets. Another Puccoon is *Lithospermum canescens,* Lehm., of the botanists. It is a rough-hairy herb of the Borage family common on the plains of the West, bearing rather large, salver-shaped orange-yellow flowers clustered at the summit of foot-high stems—

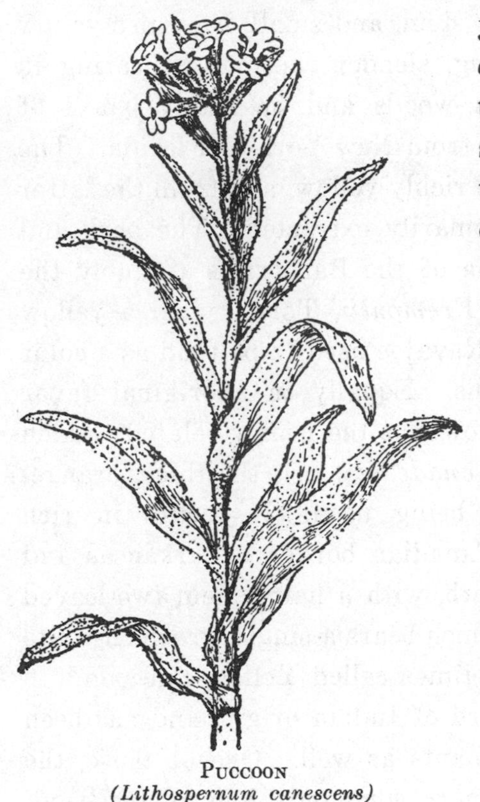

PUCCOON
(Lithospernum canescens)

Blood-root (*Sanguinaria Canadensis*), valuable as the source of a bright red dve.

(*Courtesy of the New York Botanical Gardens.*)

several from the same root. This, I believe, was the most famous of the Puccoons as an Indian color-source, a good red dye being extractable from the large red roots. The plant sometimes went among the whites by the name of Alkanet, bestowed, doubtless, because of its cousinship with the plant yielding the famous Old World dye so entitled. The Borage family, indeed, are rather rich in color juices, and some will stain the fingers even as one gathers the flowers. A red dye was also got, according to Porcher, from the fibrous roots of the Flowering Dogwood and the kindred Silky Cornel (*Cornus sericea,* L.) sometimes called Kinnikinnik. Of Kinnikinnik, more in a page or two. Another red may be extracted from the roots of the Wild Madder (*Galium tinctorium,* L.), a smooth-stemmed, perennial Bedstraw, with square stems and rather upright branches, narrow leaves in verticels usually of four, and small, 4-parted, white flowers, found in damp shade and in swampy land from Canada southward throughout much of the eastern United States. This was one of the dyes used by the northern Indians to color red the porcupine quills, which entered so largely into their decorations; and French-Canadian women, according to Kalm, employed it under the name of *tisavo jaune-rouge,* to dye cloth.

A dark blue dye Peter Kalm found in vogue among the Pennsylvania colonists, derived from the Red or Swamp Maple (*Acer rubrum,* L.), that charming

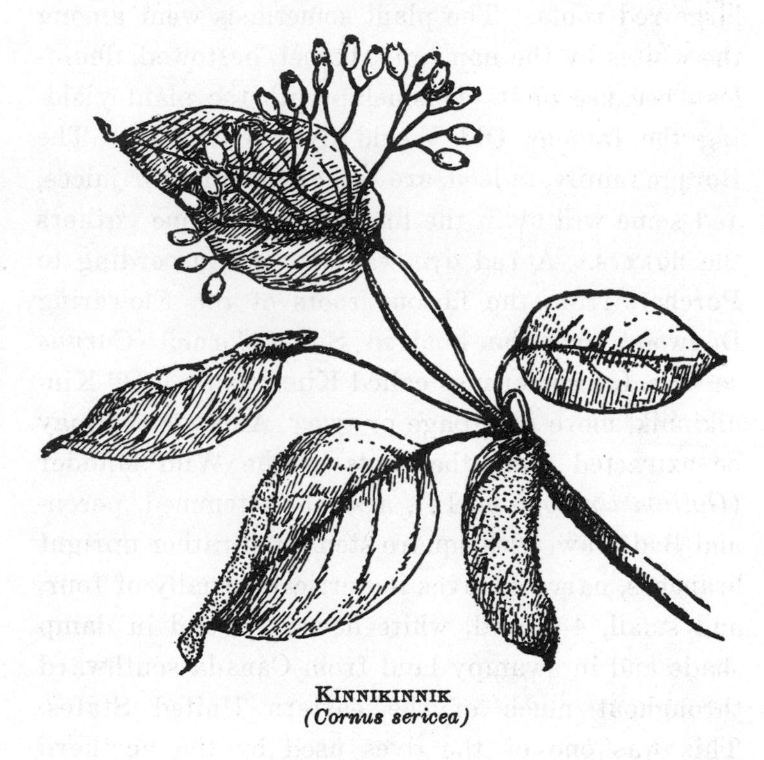

KINNIKINNIK
(Cornus sericea)

tree whose vivid blossoms, appearing before the leaves, add so much of glory to the early spring landscapes of our Atlantic seaboard. The bark, says

Kalm, is first boiled in water and before the stuff to be dyed is put into the boiler, "some copperas such as hatmakers and shoemakers use," is added. The extraction of a dark brown dye from the inner bark and the nut-rinds of the Butternut or White Walnut (*Juglans cinerea,* L.) is an old practice among country-folk, and in former times was a common method of coloring homespun woollen clothing. Civil War veterans will not yet have forgotten the butternut garments in which so many of the Confederates were clad that the term butternut became a synonym for a soldier of the South. The various species of *Alnus* or Alder, familiar shrubs (and, on the Pacific Coast, trees), contain in the bark a dye principle of value. This, in some cases, colors a brownish yellow, in others an orange. With copperas a good black may be had. Before the Indians began to use the traders' colors, alder dye was in general use among some tribes, and in the old days many an alder bush met its death through stripping by artist-squaws bent on color-getting. The bark, peeled preferably in the spring, was boiled either fresh or dried, until the water became thoroughly colored, when it was ready to receive the article to be treated.

A good Indian black has been got from the malodorous Rocky Mountain Bee-plant or Pink Spider-flower (*Cleome serrulata,* Pursh.), familiar to every traveler on our western plains, and conspicuous for its showy racemes of pink, long-stamened flowers, mingled with long-stalked, slender, outstretched seed-pods. Certain of the Pueblo Indians of New Mexico (where the plant is known among the Spanish-speaking population as *guaco*) have habitually relied upon it for the black decoration of their pottery. The plants are collected in summer, boiled down thoroughly, and the thick, black, residual fluid then allowed to dry and harden in cakes. Pieces of this are soaked in hot water, when needed for paint.[7] The desert Indians of Southern California used to obtain a yellowish-brown dye for coloring deerskins and other material from a shrubby plant of the Pea tribe, *Dalea Emoryi,* Gray, bearing small, terminal clusters of tiny pea-like flowers, staining the fingers when pinched and exhaling an odd but pleasant fragrance. The branchlets were steeped in water to release the color. Another desert dye, but black, may be had by soaking the stems of *Sueda suffrutescens,* Wats., a somewhat woody plant of the Salt-bush family, with small, dark green, fleshy

[7] Harrington, "Ethnobotany of the Tewa Indians."

leaves, found in alkaline ground from California to New Mexico.

People who have an aversion to Lady Nicotine may be interested in certain plants useful to weaken the effect of tobacco or to act as a substitute. Before the coming of the white man, the Indian smoked principally as a religious rite, as an offering of respect to superiors, or to cure disease. It was reserved for the white man to make of the practice a purely pleasurable indulgence. Moreover, the smoking material of pre-Columbian days within the territory of the present United States, was quite different from Twentieth Century commercial tobacco. There are several indigenous species of *Nicotiana,* which the aboriginal inhabitants dried and utilized, and in some instances cultivated. Their customary "smoke," however, was not pure tobacco, but a combination with other material; and this brings us again to Kinnikinnik, mentioned a little while ago. This word is an Algonkian-Indian expression signifying a mixture, and was applied by the plainsmen, trappers and settlers in the Fur Trade days to a preparation of tobacco with the dried leaves or bark of certain plants. Afterwards it came to be given to the plants themselves, the most important of which are these:

The Silky Cornel (*Cornus sericea,* L.) a shrub of wet situations, with purplish branches—these and the underleaf surfaces silky with hairs—and flattish clusters of small white flowers in early summer, succeeded in autumn by pale blue berries;

The Red-osier Dogwood (*Cornus stolonifera,* Michx.), somewhat similar to the above, but less hairy and fewer-flowered, the berries whitish, the branches smooth and brightly reddish, the plant spreading by running suckers;

The Bear-berry (*Arctostaphylos Uva-ursi,* Spreng.), a trailing, evergreen vine, with little, urn-shaped, white flowers in spring, and crimson, dryish, astringent berries in autumn, affecting rocky or sandy soil;

The Sumac, especially *Rhus glabra,* L., with smooth, pinnate leaves and smooth twigs.

In the case of the first two plants, the scraped, inner bark was the part availed of; in that of the last two, the leaves. The foliage also of Manzanita and Arrow-wood (species of *Viburnum*) sometimes found favor. The ingredients of the "smoke" were first thoroughly dried either in the sun or over a fire, and then rubbed and crumbled between the palms of the hands—whence the French *engagés'* name, *bois roulé,* applied to such smoking material. Though a portion of tobacco was usual in the make-up, it frequently was omitted—one or more of the non-narcotics being consumed alone.

When our attention is once turned to utilizing what is growing freely around us, an almost exhaustless subject of remarkable fascination has been started; and the folk of simple habits and gifted with some ingenuity find Flora a ministrant goddess of very varied gifts. There is almost nothing we can ask of her that she cannot make some sort of response to. Lovers of the curious may have napkin rings or candle-sticks from sections of the reticulated wooden skeleton of the savage Cholla Cactus; combination brushes for sweeping the floor or brushing the hair (according to the end used) from certain western grasses;[8] combs of pine-cones; buttons of acorn-cups; tooth-brushes of the Flowering Dog-

[8] One, given me by a Zuñi Indian, is a simple bunch of *Muhlenbergia pungens,* Thurb., tied about with a string, the butt-end charred to serve for the hairbrush, the other doing duty as a whisk. Harrington states that among the Tewa of New Mexico and Arizona, the plant used for this double purpose is the Mesquite-grass (*Bouteloua curtipendula,* Torr.).

wood's peeled twigs, highly recommended in old times for their whitening effect when rubbed upon the teeth.

Certain plants may even be made to yield salt, by being burned to ashes. One such is the Sweet Coltsfoot (*Petasites palmata,* Gray), a perennial herb of the Composite tribe, having large, rounded, deeply fingered leaves, all basal, white-woolly beneath and from six to ten inches broad when full grown, the whitish, fragrant flower-heads tubular or short rayed and clustered at the top of a stout, scaly stalk. The plant frequents swamps and stream borders from Massachusetts to California and far northward throughout Canada. To some Indian tribes, the ash of the Sweet Coltsfoot was their only salt. Chesnut states that the method of preparation observed by him was to roll the green leaves and stems into balls, carefully dry them, and then burn them upon a very small fire on a rock, until consumed.

Then there are adhesives. Pine pitch naturally suggests itself for this purpose; but one of the best cements for mending broken articles may be obtained from the branches of the despised Creosote bush of the Southwestern deserts (*Larrea Mexicana,* already described). This gum is not a direct vegetable exudation, but is deposited by a tiny, parasitic scale-

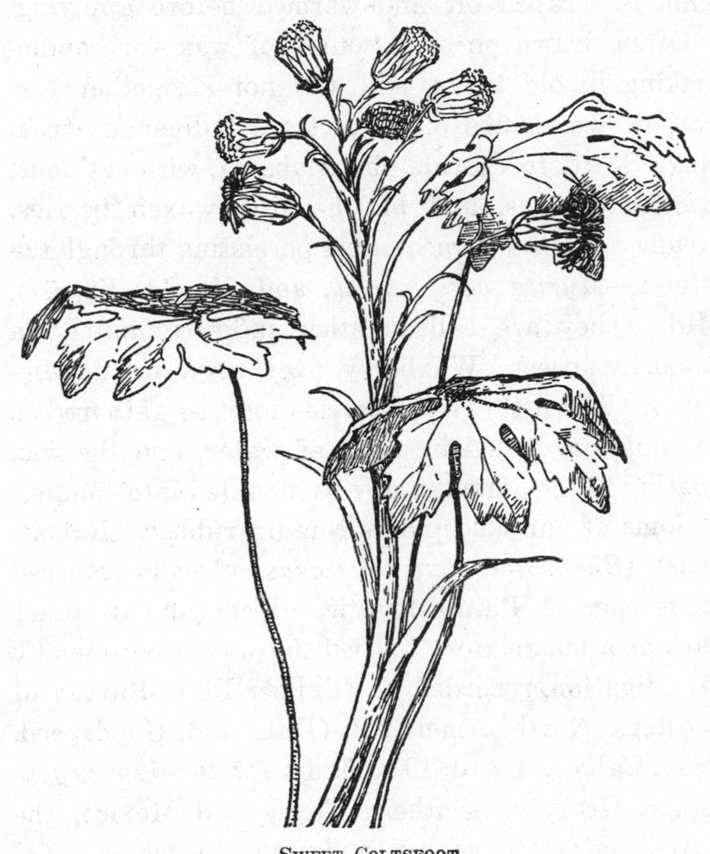

SWEET COLTSFOOT
(Petasites palmata)

insect in small reddish masses upon the twig-bark. This is scraped off and warmed before applying.

Often drawn on as a source of wax for candle-making in old times, and still not altogether forgotten, are shrubs or small trees indigenous from Nova Scotia to Florida and Alabama, with resinous, fragrant leaves, and bluish-white, waxen berries, strung upon the branches and persisting through the winter—*Myrica cerifera,* L., and *M. Carolinensis,* Mill. They are called rather indiscriminately in common speech, Waxberry, Bayberry, or Candleberry. The little round berries may be gathered in the autumn, boiled in a pot of water, and the wax floating to the surface, may be moulded into candles.

Some of our wild plants contain rubber. Rabbitbrush (*Chrysothamnus nauseosus*) already referred to, is one. A Paiute practice of chewing its inner bark as a masticatory opened the way to systematic investigation, recorded in "Rubber Plant Survey of Western North America" (Hall and Goodspeed, Univ. Calif. Press, 1919). From *Parthenium argentatum,* Gray, of southern Texas and Mexico, the Aztecs extracted rubber, and following commercial exploitation in Mexico its cultivation has been undertaken in California. Certain species of Goldenrod also have possibilities as rubber sources.

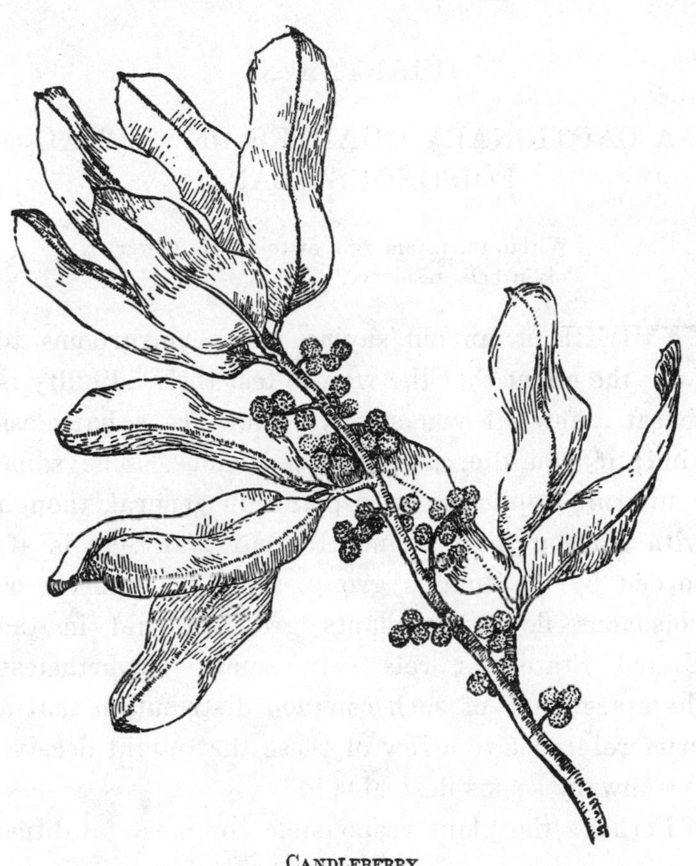

Candleberry
(Myrica Carolinensis)

CHAPTER XI

A CAUTIONARY CHAPTER ON CERTAIN POISONOUS PLANTS

"Within the infant rind of this weak flower
Poison hath residence."

THERE is an old saying about mushrooms to the effect that the way to test their edibility is to eat a few; if you survive, they are a harmless kind; if you die, they are poisonous. The same cynic rule applies to wild plants in general, though with much greater chance for survival than is afforded by the fungus group, since the number of poisonous flowering plants growing wild in the United States is relatively small. Nevertheless there are some of such common distribution that a brief reference to a few of these that might deceive the unwary seems desirable.[1]

Perhaps the plant responsible for most fatalities

[1] A useful monograph, adequately illustrated, entitled "Thirty Poisonous Plants of the United States," by V. K. Chesnut, was issued a number of years ago by the U. S. Dept. of Agriculture. as Farmers' Bulletin No. 86. Also by the same author is "Principal Poisonous Plants of the United States," Government Printing Office, 1898.

is that common toadstool appropriately called Death-cup (*Amanita phalloides*), whose resemblance to the edible Agaric or Field Mushroom (*Agaricus campestris*) causes it to be mistaken for the latter by the

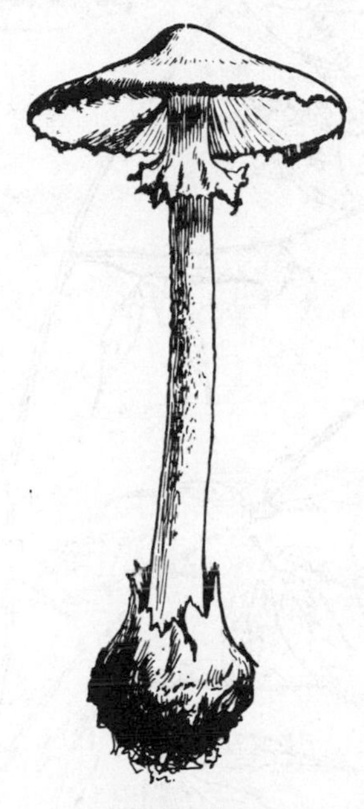

DEATH CUP
(*Amanita phalloides*)

ignorant. Any one who has not had practical instruction in differentiating edible fungi from poisonous, would best leave the fungus order religiously alone. Mushroom gathering is a business for experts.

Water Hemlock
(Cicuta maculata)

A tribe of flowering plants that includes some very dangerous members and needs to be treated with caution, is the Parsley Family—the scientists' *Umbelliferae.* To this order belongs the Water Hemlock or Cowbane (*Cicuta maculata,* L.), a perennial of marshy grounds and stream borders from the Atlantic coast westward to the confines of the Rocky Mountains. It grows from three to six feet high, with stout, erect stems blotched or streaked longitudinally with purple, and ample, compound leaves the segments of which are usually two to three inches long, lance-shaped and toothed. A peculiarity of the foliage is the veining—the veins apparently ending within the notches instead of extending to the tips of the teeth. The small white flowers, appearing in summer, are borne at the branch end in compound, long-stalked umbels, after the manner of parsley blossoms. All parts of the plant are poisonous if eaten, producing nausea and convulsions, the fleshy, tuberous roots being especially harmful. These are said to possess an agreeable, aromatic taste, and as they are often found exposed through the wearing away of the surrounding earth in freshets, they constitute a menace to inquisitive children and browsing cattle. Death results from eating them. On the Pacific coast two or three species of Water Hemlock occur, also inhabiting marshy places, and all are possessed of the same deadly properties.

The famous Poison Hemlock of Greek history and Macbeth's witches (*Conium maculatum,* L.)—the basis of the death potion of Socrates—is also a member of the Parsley family, native to Europe and Asia but now extensively naturalized in the United States in waste grounds on both sides of the continent. It is a smooth, hollow-stemmed, much branched, bluish-green biennial, sometimes as high as a tall man, but usually much lower, with large, coarsely dissected leaves, the leaf-stalks dilated at the base and sheathing. The stems are often spotted with dark purple. The small white flowers appear in June in compound, many-rayed umbels. The poisonous principle—an alkaloid called conia or conine—is permanently resident in the seeds and only temporarily in other parts of the plant. According to Chesnut, the root is nearly harmless in March, but dangerous if consumed afterwards, and the leaves become poisonous at the time of flowering. The effect of the poison is a general paralysis of the system until death. A drug, conium, prepared from the plant, is a powerful sedative and has been used medicinally as a substitute for opium.[2]

[2] One wonders why hemlock, which we associate with a forest

Butternut (*Juglans cinerea*). The bark is the source of a dye used for the uniforms of Confederate soldiers during the Civil War. (See page 227.)

(*Courtesy of the Bureau of American Ethnology.*)

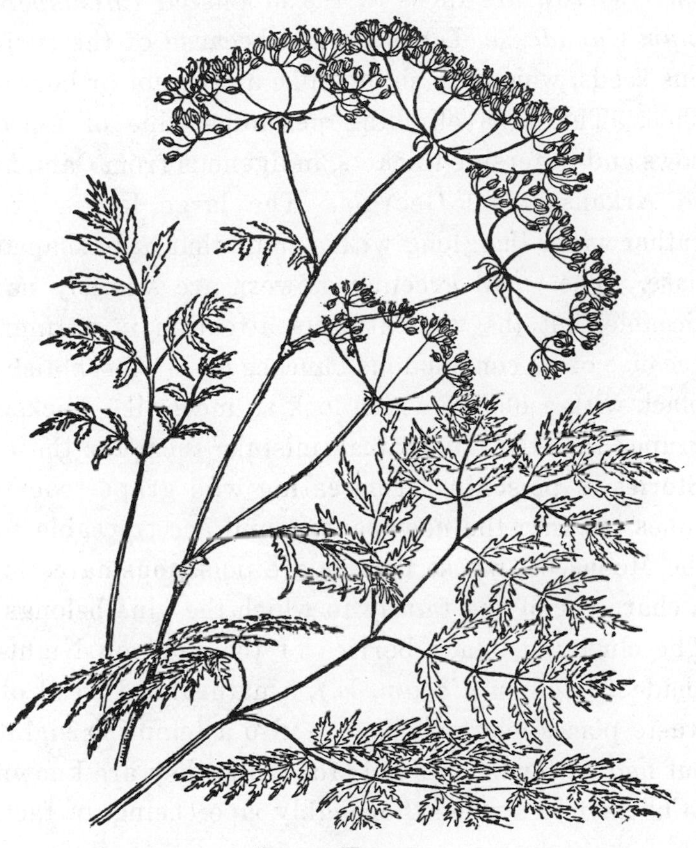

Poison Hemlock
(Conium maculatum)

Noxious berries that sometimes tempt children to their sorrow are those of the Moonseed (*Menispermum Canadense*, L.), so called because of the curious seeds, which are shaped like a crescent or horseshoe. This is a climbing perennial vine of fence rows and waterside thickets, indigenous from Canada to Arkansas and Georgia. The large leaves are rather wider than long with a somewhat heartshaped base. The small greenish flowers are scarcely noticeable, but the vine attracts attention in autumn because of its conspicuous bunches of berries, bluish-black with a bloom, which look so much like chicken grapes that the novice may mistake them for these. Stories of poisoning from eating wild grapes sometimes get into the newspapers, and are traceable to the Moonseed, whose berries are poisonous-narcotic, a character of the family to which the vine belongs. The clustered, black berries of the common Nightshade (*Solanum nigrum*, L.), a naturalized weed of waste places everywhere, are also a tempting sight, but had better be avoided; for while they are known to be harmless when thoroughly ripe (being, in fact,

tree, should be applied to an herb. According to Prior in "Popular Names of British Plants," the term was originally given in England to any of the *Umbelliferae*—the word being degenerate Anglo-Saxon meaning "straw plant," because of the dry, hollow stalks that remain after flowering.

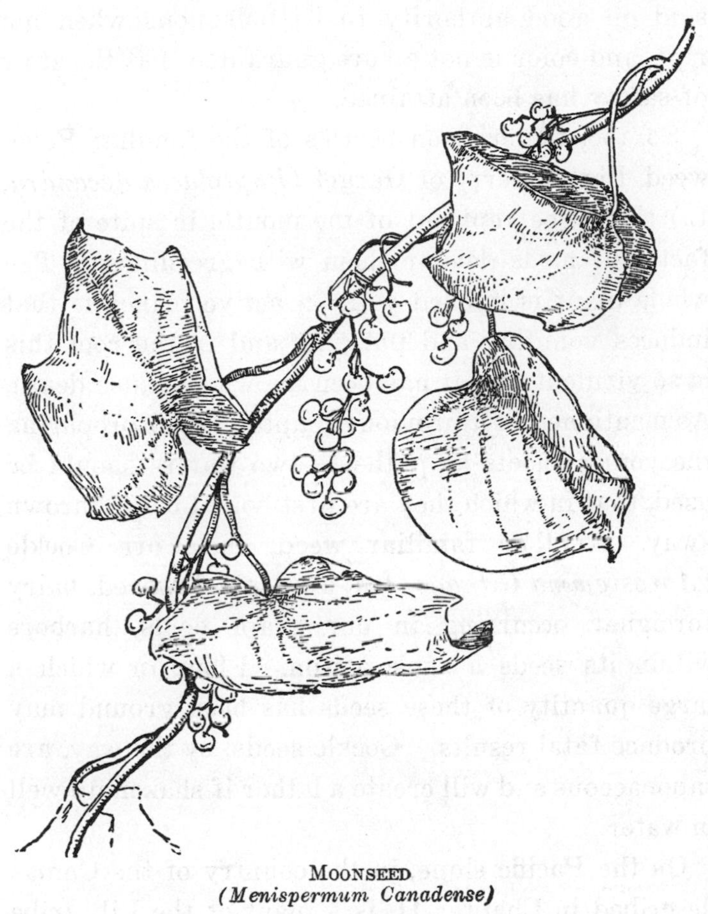

MOONSEED
(Menispermum Canadense)

used for pie-making in parts of the West), they are said on good authority to be poisonous when not ripe, and color is not a sure guarantee that the state of safety has been attained.

So, too, the crimson berries of the familiar Poke-weed, Pigeon-berry or Garget (*Phytolacca decandra*, L.) should be kept out of the mouth, in spite of the fact that birds devour them with greediness. The whole plant is imbued with an active principle that induces vomiting and purging, and in the root this is so virulent that it has been known to cause death. As mentioned in a previous chapter, when preparing the young shoots as potherbs two waters should be used, that in which they are first boiled being thrown away. Another familiar weed, the Corn Cockle (*Agrostemma Githago*, L.), a purple flowered, hairy foreigner occurring in our grain fields, harbors within its seeds a rank poison. Flour in which a large quantity of these seeds has been ground may produce fatal results. Cockle seeds, by the way, are saponaceous and will create a lather if shaken up well in water.

On the Pacific slope, in the country of the Camas described in Chapter II, is a plant of the Lily tribe in general appearance resembling Camas but with a bulb that is poisonous. It is realistically known as

Death Camas, and also as White Camas and Lobelia. It haunts damp meadows and streamsides, and is in botanical parlance *Zygadenus venenosus,* Wats. The white flowers serve to distinguish it from the blue Camas, which otherwise it strongly simulates. The effect of eating the Zygadenus bulb is a profound nausea accompanied by vomiting. Mr. F. V. Coville records a crafty practice of the Klamath medicine men, who would sometimes make a mixture of tobacco, dried iris root and Death Camas, and give it to a person in order to nauseate him. Then they would charge the victim a fee to make him well again!

A poison unsuspected by most of us resides in the leaves of that beautiful evergreen shrub, the American Laurel or Calico-bush (*Kalmia latifolia,* L.), which glorifies with its white and pink bloom the spring thickets of the Atlantic seaboard. Man has little occasion to put these leaves in his mouth, but the ill effect upon cattle and sheep has been often reported. A like offender is the Laurel's little red-flowered cousin, the Sheep-Laurel or Lambkill (*K. angustifolia,* L.). Stock may also suffer fatally from eating the wilted foliage of the Wild Black Cherry (*Prunus serotina,* a tree already described, with clusters of edible, small, black, somewhat astringent fruit). The most dreaded of cattle-poisons, however, particularly on the Western ranges, is probably the so-called Loco-weed, a term applied to several species of Astragalus—especially *A. mollissimus,* Torr., distinguished by purple flowers and densely hairy foliage. The genus is of the

LOCO-WEED
(Astragalus mollissimus)

Pea family, and is a very large one, widely distributed. There are nearly two hundred American species, mostly western—herbaceous plants with odd-pinnate leaves, spikes or racemes of usually small, narrow flowers generally produced from the leaf-axils, the seed pods mostly bladdery or swollen. These, when dry, have a habit of rustling noticeably in a passing breeze, whence another common name, Rattleweed. Astragalus is often abundant where horses and cattle graze, and certain species have been found to create serious trouble with animals that eat the herbage. They become afflicted with a sort of insanity, or as the Westerners say, they are "locoed,"[3] the victims of a slow poisoning. The eyesight grows defective, the movements are spasmodic and irrational, then sluggish and feeble, the coat becomes disheveled and dull of color, emaciation sets in, and finally after a few months or it may be a year or two, death comes. It was at one time thought that the poisoning was not of the plant itself but due to the presence of the metal barium which the plant drew into its system from the soil, but this theory is now abandoned.

A dangerously poisonous weed is the Jimson or Thorn-apple (*Datura Stramonium*, L.), whose large

[3] Spanish *loco*, crazy, foolish.

funnel-shaped, white or violet flowers and thorny seed-vessels adorning ill-smelling, branching plants,

JIMSON-WEED
(Datura Stramonium)

are familiar sights in fields and waste grounds from the Mississippi eastward and from Canada to the Gulf. The whole plant and particularly the seeds are possessed of a virulent narcotic poison, which taken into the human body produces vertigo, nausea, delirium and a general anarchy of the nervous system. In that quaint old work, "History and Present State of Virginia" (1705), by Robert Beverly, the author gives a curious account of what happened to some soldiers who made a boiled dish of the early shoots of the plant, supposing them to be edible potherbs. "Some of them eat plentifully of it," writes Master Beverly, "the Effect of which was a very pleasant Comedy; for they turn'd natural Fools upon it for several Days: One would blow up a Feather in the Air; another would dart Straws at it with much Fury; another, stark naked, was sitting in a Corner, like a Monkey, grinning and making mows at them; a Fourth would fondly kiss and paw his Companions and snear in their Faces with a Countenance more antick than any Dutch Droll. . . . A thousand such simple Tricks they play'd, and after Eleven Days, return'd to themselves again, not remembering anything that had pass'd."[4]

There are several species of Datura indigenous within our limits, all resembling one another in general look and all poisonous. On the Pacific Slope,

[4] Beverly calls the plant James Town weed, which seems to have been the original term, now corrupted to Jimson.

the commonest species is *D. meteloides*, DC., called *toloache* by Mexicans and Indians. This, like several species of Spanish America, has played a noteworthy part in the ceremonial life of our aborigines. An infusion of the plant was customarily administered in certain rites, as those of puberty; and it was a drug commonly resorted to by medicine men to induce a hypnotic state or a condition evocative of prophecy. Only a little while ago a California Indian expressed to me his faith in the power of *toloache* to unravel mysteries and reveal the whereabouts of lost animals. The likelihood of death from overindulgence makes its employment risky, and it is nowadays comparatively neglected. Among the New Mexico Zuñis, the blossom of this Datura is a sacred flower, and a representation of it figures as an adornment of the women in some of their dances. Mrs. Stevenson in her "Ethnobotany of the Zuñi Indians,"[5] records a legend about this flower worthy of Ovid. It seems that long, long ago while the Zuñis still dwelt in the underworld, a boy and a girl, brother and sister, found a way up into this world of light, and would take long walks upon the earth, wearing upon their heads Datura flowers. And so they learned many wonderful things, and had many

[5] **30th Ann. Rept. Bureau of American Ethnology.**

interesting adventures. One day they met the Divine Ones, the Twin Sons of the Sun Father, to whom, child-like, they prattled of what they had found out—how they could make people sleep and see ghosts, and how they could make others walk about and see who it was that had stolen something. Thereupon the Divine Ones decided that this little couple knew altogether too much, and should be made away with. So they caused the brother and sister to disappear into the earth forever; and where they sank down flowers sprang up, the counterpart of those that the children had worn upon their heads. The gods called the flowers by the name of the boy, *Aneglakya*; and by that term the Zuñis know them to this day, for the flowers had many children and we find them throughout the land.

In western Texas and southern New Mexico, ranging across the frontier down into Old Mexico, there grows a handsome shrub of the Pea family, with glossy, odd-pinnate, evergreen leaves of leathery texture, and one sided racemes of papilionaceous, violet-colored flowers, succeeded by long pods that contain about half a dozen large scarlet bean-like seeds apiece. This is the Red Bean, Mescal Bean, or as the Spanish-speaking population call it, *Frijolillo,* which means the "little pink bean." To botanists it is *Broussonetia secundiflora,* Ort., or *Sophora secundiflora,* Lag. The seeds contain a narcotic poison that makes them dangerous particularly to children, who are likely to be attracted by the brilliant color. The crushed seeds have been used from very early times by the Indians, who, it is reported, could make themselves deliriously drunk on half a bean, and sleep two or three days on top of it, while a whole bean would kill a man. Among some tribes, as the Iowas, there were religious rites connected with the Red Bean, and a society was founded upon it.

To-day one hears little of the Red Bean Society, but the cult of another dangerous vegetable poison of the Southwest is still active. This is the so-called Sacred Mushroom, Mescal-button, Dry Whisky, Peyote, or *Raiz diabólica* (devil's root)—names given in common speech to a small cactus, *Lophophora Williamsii,* whose use has become a rather desolating factor among the present-day Reservation Indians of the United States. Some of these, it appears, maintain a regularly organized association called the Sacred Peyote Society with a form of baptism "in the name of the Father, and the Son and the Holy Ghost," the Holy Ghost being Peyote![6]

[6] Quoted by W. E. Safford, "Narcotic Plants and Stimulants of

Indian woman preparing squaw-weed (*Rhus trilobata*) for basket making. (See page 154.)

The cactus is indigenous to the arid regions bordering on the lower Rio Grande both in the United States and Mexico. It resembles a carrot in shape, and the entire plant, except about an inch at the top, grows underground. This top is flat and round, two to three inches across, and wrinkled with radiating ribs. There are no spines but numerous tufts of silky hairs, amid which pink blossoms are borne in season. The chemical properties embrace three alkaloids whose effect is powerfully narcotic and delirant, in some respects resembling opium. Lumholtz, in his "Unknown Mexico," gives an inter-

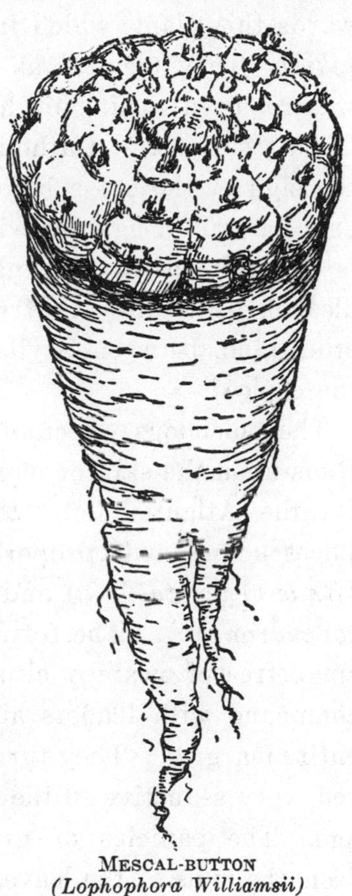

MESCAL-BUTTON
(Lophophora Williamsii)

the Ancient Americans," in Ann. Rept. Smithsonian Institution, 1916.

esting account of the superstitious reverence accorded by the Tarahumar Indians of Chihuahua towards this plant, which in their language is called *hikuli.* They treat it as a divinity and Lumholtz was required to lift his hat in the presence of the dried "buttons." Catholicized Tarahumares make the sign of the cross before it; and it is regarded as a safeguard against witches and ill fortune. It is claimed that its use takes away the craving for alcohol, which may be true; but it substitutes another, and, between Scylla and Charybdis, what is the choice?

The poisonous effect of a few native species of Rhus upon the skin of many persons is well known. On the Atlantic slope the species whose caustic juices possess this property are the Swamp Sumac (*Rhus venenata,* DC.) and the Poison Ivy (*R. Toxicodendron,* L.). The former is a graceful shrub or small tree of swampy situations, the smooth leaves compound with leaflets abruptly pointed and with entire margins. They turn in the autumn a brilliant red, very seductive to the gatherers of autumn foliage. The panicles of greenish flowers, produced from the axils of the leaves, are followed by grayish white berries. The plant is also called Poison Sumac and, less correctly, Poison Elder. The

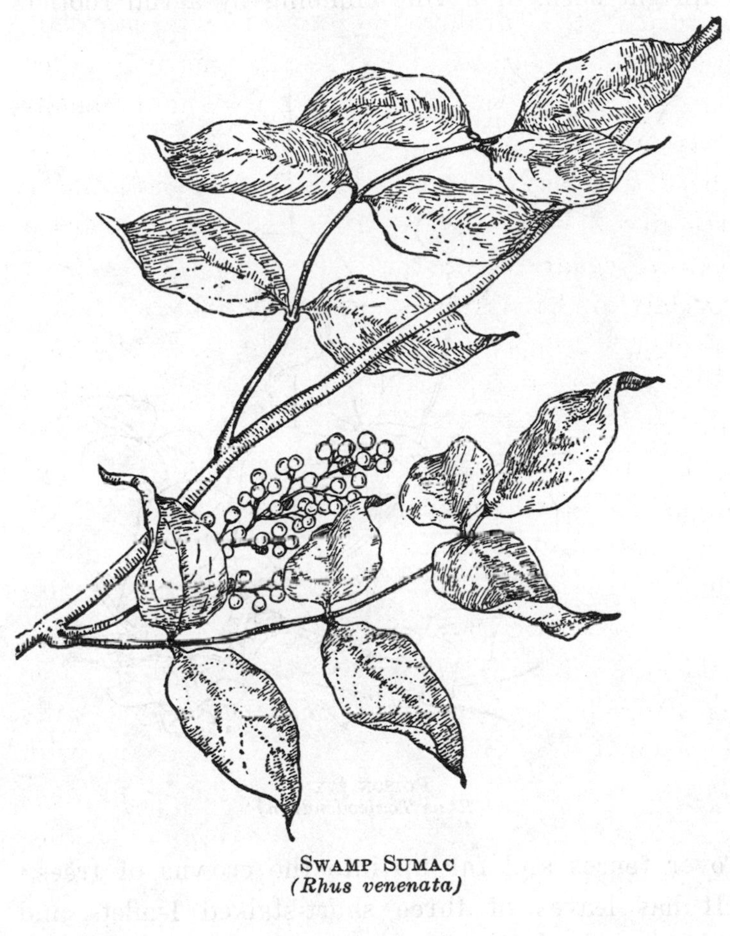

Swamp Sumac
(Rhus venenata)

Poison Ivy is very variable in habit, either a low, upright bush, or a vine climbing by aërial rootlets

POISON IVY
(Rhus Toxicodendron)

over fences and far up into the crowns of trees.[7] It has leaves of three short-stalked leaflets, and

[7] Some botanists prefer to treat Poison Ivy as of two species—the climber being designated *Rhus radicans.*

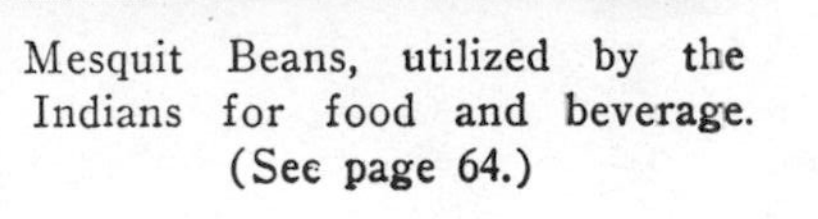

Mesquit Beans, utilized by the Indians for food and beverage. (See page 64.)

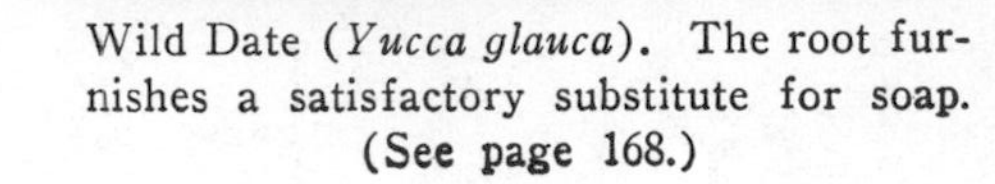

Wild Date (*Yucca glauca*). The root furnishes a satisfactory substitute for soap. (See page 168.)

flowers and fruit like those of the Swamp Sumac. This 3-leaflet arrangement serves to distinguish the plant from the harmless but somewhat similar looking Virginia Creeper or American Ivy, which has leaves of five parts. On the Pacific Slope, the representative poisonous Rhus is *R. diversiloba,* T. & G., commonly called Poison Oak. It is in general appearance like the eastern Poison Ivy, either bushy or climbing, but the leaflets are variously lobed and toothed, suggesting an oak. Popular remedies for rhus poisoning are legion and include lotions made from a variety of native plants, including Bloodroot, Virginia Snake-root, Jimson-weed, Manzanita leaves, Grindelia, Amole juice, and so on. It is questionable, however, if their value be not in the same category with the herb doctors' rattlesnake "cures."[8] The historian Bancroft records that a Spanish expedition in the Southwest early in the eighteenth century, under Governor Valverde, suffered greatly from Poison Oak and found relief by chewing chocolate and applying the saliva to the eruption. Rather a pleasing remedy, on the whole, one would fancy; and I am glad to think of those old campaigners in the desert having that little taste of sweet in the bitterness of their lot.

8 "Rhus Dermatitis" by James B. McNair, Univ. of Chicago Press, 1923, treats the subject exhaustively from a chemist's standpoint.

REGIONAL INDEX

(For Page Numbers see General Index.)

The notation (A) after a plant indicates that it is found only in the Atlantic States. The notation (W) after a plant indicates that it is found only west of the Atlantic States.

East of the Rocky Mountains (including Middle and Eastern Canada)

Food Plants:

Edible Roots and Tubers:

Arrowhead (Sagittaria variabilis)
Chufa (Cyperus esculenta)
Golden Club (Orontium aquaticum) A
Groundnut (Apios tuberosa)
Indian Bread-root (Psoralea esculenta) W
Jack-in-the-Pulpit (Arisaema triphyllum)
Jerusalem Artichoke (Helianthus tuberosus) W
Man-of-the-earth (Ipomoea pandurata)
Tobacco-root (Valeriana edulis)
Virginia Tuckaho (Peltandra Virginica)
Water Chinquapin (Nelumbo lutea)
Wild Onion (Allium tricoccum)

Edible Seeds:

Beechnut (Fagus Americana)
Chestnut (Castanea dentata)
Chinquapin (Castanea pumila)
Golden Club (Orontium aquaticum)
Groundnut (Apios tuberosa)
Hickory (Hicoria sp.)

Hog Peanut (Amphicarpaea monoica)
Sunflower (Helianthus annuus sp.) w
Walnut (Juglans sp.)
Water Chinquapin (Nelumbo lutea)
Wild Rice (Zizania aquatica)

Edible Fruits and Berries:

Barberry (Berberis sp.)
Blackberry (Rubus sp.)
Buffalo-berry (Shepherdia argentea) w
Cranberry (Oxycoccus sp.)
Currant (Ribes sp.)
Gooseberry (Ribes sp.)
Grape (Vitis sp.)
Ground Cherry (Physalis sp.)
Hawthorn (Crataegus sp.)
Huckleberry (Vaccinium sp.)
May Apple (Podophyllum peltatum)
Mulberry (Morus rubra)
Papaw (Asimina triloba)
Persimmon (Diospyros Virginica)
Raspberry (Rubus sp.)
Service-berry (Amelanchier sp.)
Strawberry (Fragaria sp.)
Teaberry (Gaultheria procumbens)

Edible Stems or Leaves:

Bracken (Pteris aquilina)
Chicory (Cichorium Intybus)
Dandelion (Taraxacum officinale)
Dock (Rumex crispus)
Lamb's quarters (Chenopodium album)
Milkweed (Asclepias sp.)
Nettle (Urtica dioica)
Pokeweed (Phytolacca decandra)
Purslane (Portulaca oleracea)
Water-cress (Nasturtium officinale)
Winter Cress (Barbarea vulgaris)

REGIONAL INDEX

REGIONAL INDEX

REGIONAL INDEX

Edible Fruits:
May-pop (Passiflora incarnata)
Summer Haw (Crataegus flava)

Edible Stems or Leaves:
Cabbage Palmetto (Sabal Palmetto)
Scurvy Grass (Barbarea praecox)

BEVERAGE PLANTS:
Cassena (Ilex vomitoria)

SOAP-PLANTS:
Soap-berry (Sapindus sp.)
Southern Buckeye (Aesculus Pavia)

DYE-PLANTS:
Kentucky Yellow-wood (Cladastris tinctoria)
Shrub-Yellow-root (Xanthorrhiza apiifolia)

THE PACIFIC SLOPE

FOOD PLANTS:

Edible Roots and Tubers:
Arrowhead (Sagittaria variabilis)
Biscuit-root (Peucedanum sp.)
Bitter-root (Lewisia rediviva)
Camas (Camassia esculenta)
Chufa (Cyperus esculentus)
Harvest Brodiaea (Brodiaea grandiflora)
Indian Potatoes (Calochortus sp., Camassia sp., Brodiaea sp., etc.)
Sego Lily (Calochortus Nuttallii)
Tule (Scirpus lacustris)
Wild Anise (Carum Kelloggii)
Wild Onion (Brodiaea capitata)
Yamp (Carum Gairdneri)

Edible Seeds:
Buckeye (Aesculus Californicus)
Chia (Salvia sp.)

Chinquapin (Castanopsis chrysophylla)
Goosefoot (Chenopodium sp.)
Islay (Prunus ilicifolia)
Oak (Quercus sp.)
Pine (Pinus sp.)
Pond-lily (Nymphaea polysepala)
Sunflower (Helianthus annuus)
Tarweed (Madia sativa)
Walnut (Juglans Californica)
White Sage (Audibertia polystachya)
Wild Oats (Avena fatua)
Wild Wheat (Elymus triticoides)

Edible Fruits and Berries:

Black Haw (Crataegus Douglasii)
Buckthorn (Rhamnus crocea)
Cranberry (Oxycoccus sp.)
Currant (Ribes aureum)
Grape (Vitis Californica)
Huckleberry (Vaccinium sp.)
Manzanita (Arctostaphylos sp.)
Oregon Grape (Berberis aquifolium)
Raspberry (Salmon-berry, Thimbleberry) (Rubus sp.)
Salal (Gaultheria Shallon)
Service-berry (Amelanchier sp.)
Strawberry (Fragaria sp.)
Tuna (Opuntia sp.)

Edible Stems or Leaves:

Bracken (Pteris aquilina)
Clover (Trifolium)
Miner's Lettuce (Montia perfoliata)
Purslane (Portulaca oleracea)
Red Maids (Calandrinia caulescens Menziesii)
Water-cress (Nasturtium officinale)
Wild Pie-plant (Rumex hymenosepalus)

BEVERAGE PLANTS:

Chia (Salvia sp.)

Douglas Spruce (Pseudotsuga taxifolia)
Lemonade-berry (Rhus sp.)
Manzanita (Arctostaphylos sp.)
Yerba buena (Micromeria Douglasii)

SOAP-PLANTS:

Amole (Chloragalum pomeridianum)
Mock Orange (Cucurbita foetidissima)
Soap-plant (Chlorogalum pomeridianum)
Soap-root (Chenopodium Californicum)
Wild Lilac (Ceanothus sp.)

MEDICINAL PLANTS:

California Laurel (Umbellularia Californica)
Canchalagua (Erythraea venusta)
Cascara sagrada (Rhamnus Californica)
Gum-plant (Grindelia sp.)
Hoar-hound (Marrubium vulgare)
Mastransia (Stachys Californica)
Mustard (Brassica sp.)
Quinine-bush (Garrya elliptica)
Western Dogwood (Cornus Nuttallii)
Yarrow (Achillea Millifolium)
Yerba mansa (Anemopsis Californica)
Yerba santa (Eriodictyon glutinosum)

FISH POISONS:

Soap-root (Chloragalum pomeridianum)
Turkey Mullein (Croton setigerus)

FIBER PLANTS:

Indian Hemp (Apocynum cannabinum)
Milkweed (Asclepias eriocarpa)
Psoralea (Psoralea macrostachya)

DYE PLANTS:

Alder (Alnus sp.)
Sunflower (Helianthus annuus)

REGIONAL INDEX

The Southwest (Mainly in Arid Regions)

ADDENDUM TO REGIONAL INDEX

EAST OF THE ROCKY MOUNTAINS

FOOD PLANTS:

Edible Fruits and Berries:

Florida Palmetto (Sabal Palmetto) A

Edible Stems or Leaves:

Cinnamon and Interrupted Ferns (Osmunda)
Jewel-weed (Impatiens fulva)
Rock-tripe (Umbilicaria)

BEVERAGE PLANTS:

Creeping Snowberry (Chiogenes hispidula)

MEDICINAL PLANTS:

Goldenrod (Solidago odora) A

THE PACIFIC SLOPE AND THE SOUTHWEST

FOOD PLANTS:

Edible Roots and Tubers:

Kooyah or Tobacco-root (Valeriana edulis)
Thistle (Cirsium edule)

Edible Fruits or Berries:

California Holly or Toyon (Photinia arbutifolia)
Wild Rose (Rosa sp.)

Edible Stems or Leaves:

Ice-plant (Mesembryanthemum crystallinum)

BEVERAGE PLANT:

Bird's-foot Fern (Pellaea ornithopus)

MEDICINAL PLANTS:

Bird's-foot Fern (Pellaea ornithopus)
Goldenrod (Solidago californica)

RUBBER PLANTS:

Guayule (Parthenium argentatum)
Rabbit-brush (Chrysothamnus nauseosus)

INDEX

INDEX

INDEX

INDEX

INDEX

INDEX

INDEX

ADDENDUM TO INDEX

A CATALOGUE OF SELECTED DOVER BOOKS IN ALL FIELDS OF INTEREST

A CATALOGUE OF SELECTED DOVER BOOKS IN ALL FIELDS OF INTEREST

CELESTIAL OBJECTS FOR COMMON TELESCOPES, T. W. Webb. The most used book in amateur astronomy: inestimable aid for locating and identifying nearly 4,000 celestial objects. Edited, updated by Margaret W. Mayall. 77 illustrations. Total of 645pp. 5⅜ x 8½.
20917-2, 20918-0 Pa., Two-vol. set $9.00

HISTORICAL STUDIES IN THE LANGUAGE OF CHEMISTRY, M. P. Crosland. The important part language has played in the development of chemistry from the symbolism of alchemy to the adoption of systematic nomenclature in 1892. ". . . wholeheartedly recommended,"—Science. 15 illustrations. 416pp. of text. 5⅝ x 8¼. 63702-6 Pa. $6.00

BURNHAM'S CELESTIAL HANDBOOK, Robert Burnham, Jr. Thorough, readable guide to the stars beyond our solar system. Exhaustive treatment, fully illustrated. Breakdown is alphabetical by constellation: Andromeda to Cetus in Vol. 1; Chamaeleon to Orion in Vol. 2; and Pavo to Vulpecula in Vol. 3. Hundreds of illustrations. Total of about 2000pp. 6⅛ x 9¼.
23567-X, 23568-8, 23673-0 Pa., Three-vol. set $26.85

THEORY OF WING SECTIONS: INCLUDING A SUMMARY OF AIRFOIL DATA, Ira H. Abbott and A. E. von Doenhoff. Concise compilation of subatomic aerodynamic characteristics of modern NASA wing sections, plus description of theory. 350pp. of tables. 693pp. 5⅜ x 8½.
60586-8 Pa. $7.00

DE RE METALLICA, Georgius Agricola. Translated by Herbert C. Hoover and Lou H. Hoover. The famous Hoover translation of greatest treatise on technological chemistry, engineering, geology, mining of early modern times (1556). All 289 original woodcuts. 638pp. 6¾ x 11.
60006-8 Clothbd. $17.50

THE ORIGIN OF CONTINENTS AND OCEANS, Alfred Wegener. One of the most influential, most controversial books in science, the classic statement for continental drift. Full 1966 translation of Wegener's final (1929) version. 64 illustrations. 246pp. 5⅜ x 8½. 61708-4 Pa. $3.00

THE PRINCIPLES OF PSYCHOLOGY, William James. Famous long course complete, unabridged. Stream of thought, time perception, memory, experimental methods; great work decades ahead of its time. Still valid, useful; read in many classes. 94 figures. Total of 1391pp. 5⅜ x 8½.
20381-6, 20382-4 Pa., Two-vol. set $13.00

YUCATAN BEFORE AND AFTER THE CONQUEST, Diego de Landa. First English translation of basic book in Maya studies, the only significant account of Yucatan written in the early post-Conquest era. Translated by distinguished Maya scholar William Gates. Appendices, introduction, 4 maps and over 120 illustrations added by translator. 162pp. 5⅜ x 8½.
23622-6 Pa. $3.00

THE MALAY ARCHIPELAGO, Alfred R. Wallace. Spirited travel account by one of founders of modern biology. Touches on zoology, botany, ethnography, geography, and geology. 62 illustrations, maps. 515pp. 5⅜ x 8½.
20187-2 Pa. $6.95

THE DISCOVERY OF THE TOMB OF TUTANKHAMEN, Howard Carter, A. C. Mace. Accompany Carter in the thrill of discovery, as ruined passage suddenly reveals unique, untouched, fabulously rich tomb. Fascinating account, with 106 illustrations. New introduction by J. M. White. Total of 382pp. 5⅜ x 8½. (Available in U.S. only) 23500-9 Pa. $4.00

THE WORLD'S GREATEST SPEECHES, edited by Lewis Copeland and Lawrence W. Lamm. Vast collection of 278 speeches from Greeks up to present. Powerful and effective models; unique look at history. Revised to 1970. Indices. 842pp. 5⅜ x 8½. 20468-5 Pa. $8.95

THE 100 GREATEST ADVERTISEMENTS, Julian Watkins. The priceless ingredient; His master's voice; 99 44/100% pure; over 100 others. How they were written, their impact, etc. Remarkable record. 130 illustrations. 233pp. 7⅞ x 10 3/5. 20540-1 Pa. $5.00

CRUICKSHANK PRINTS FOR HAND COLORING, George Cruickshank. 18 illustrations, one side of a page, on fine-quality paper suitable for watercolors. Caricatures of people in society (c. 1820) full of trenchant wit. Very large format. 32pp. 11 x 16. 23684-6 Pa. $5.00

THIRTY-TWO COLOR POSTCARDS OF TWENTIETH-CENTURY AMERICAN ART, Whitney Museum of American Art. Reproduced in full color in postcard form are 31 art works and one shot of the museum. Calder, Hopper, Rauschenberg, others. Detachable. 16pp. 8¼ x 11.
23629-3 Pa. $2.50

MUSIC OF THE SPHERES: THE MATERIAL UNIVERSE FROM ATOM TO QUASAR SIMPLY EXPLAINED, Guy Murchie. Planets, stars, geology, atoms, radiation, relativity, quantum theory, light, antimatter, similar topics. 319 figures. 664pp. 5⅜ x 8½.
21809-0, 21810-4 Pa., Two-vol. set $10.00

EINSTEIN'S THEORY OF RELATIVITY, Max Born. Finest semi-technical account; covers Einstein, Lorentz, Minkowski, and others, with much detail, much explanation of ideas and math not readily available elsewhere on this level. For student, non-specialist. 376pp. 5⅜ x 8½.
60769-0 Pa. $4.00

TONE POEMS, SERIES II: TILL EULENSPIEGELS LUSTIGE STREICHE, ALSO SPRACH ZARATHUSTRA, AND EIN HELDENLEBEN, Richard Strauss. Three important orchestral works, including very popular *Till Eulenspiegel's Marry Pranks,* reproduced in full score from original editions. Study score. 315pp. 9⅜ x 12¼. (Available in U.S. only)
23755-9 Pa. $7.50

TONE POEMS, SERIES I: DON JUAN, TOD UND VERKLARUNG AND DON QUIXOTE, Richard Strauss. Three of the most often performed and recorded works in entire orchestral repertoire, reproduced in full score from original editions. Study score. 286pp. 9⅜ x 12¼. (Available in U.S. only) 23754-0 Pa. $7.50

11 LATE STRING QUARTETS, Franz Joseph Haydn. The form which Haydn defined and "brought to perfection." *(Grove's).* 11 string quartets in complete score, his last and his best. The first in a projected series of the complete Haydn string quartets. Reliable modern Eulenberg edition, otherwise difficult to obtain. 320pp. 8⅜ x 11¼. (Available in U.S. only)
23753-2 Pa. $6.95

FOURTH, FIFTH AND SIXTH SYMPHONIES IN FULL SCORE, Peter Ilyitch Tchaikovsky. Complete orchestral scores of Symphony No. 4 in F Minor, Op. 36; Symphony No. 5 in E Minor, Op. 64; Symphony No. 6 in B Minor, "Pathetique," Op. 74. Bretikopf & Hartel eds. Study score. 480pp. 9⅜ x 12¼. 23861-X Pa. $10.95

THE MARRIAGE OF FIGARO: COMPLETE SCORE, Wolfgang A. Mozart. Finest comic opera ever written. Full score, not to be confused with piano renderings. Peters edition. Study score. 448pp. 9⅜ x 12¼. (Available in U.S. only) 23751-6 Pa. $11.95

"IMAGE" ON THE ART AND EVOLUTION OF THE FILM, edited by Marshall Deutelbaum. Pioneering book brings together for first time 38 groundbreaking articles on early silent films from *Image* and 263 illustrations newly shot from rare prints in the collection of the International Museum of Photography. A landmark work. Index. 256pp. 8¼ x 11.
23777-X Pa. $8.95

AROUND-THE-WORLD COOKY BOOK, Lois Lintner Sumption and Marguerite Lintner Ashbrook. 373 cooky and frosting recipes from 28 countries (America, Austria, China, Russia, Italy, etc.) include Viennese kisses, rice wafers, London strips, lady fingers, hony, sugar spice, maple cookies, etc. Clear instructions. All tested. 38 drawings. 182pp. 5⅜ x 8.
23802-4 Pa. $2.50

THE ART NOUVEAU STYLE, edited by Roberta Waddell. 579 rare photographs, not available elsewhere, of works in jewelry, metalwork, glass, ceramics, textiles, architecture and furniture by 175 artists—Mucha, Seguy, Lalique, Tiffany, Gaudin, Hohlwein, Saarinen, and many others. 288pp. 8⅜ x 11¼. 23515-7 Pa. $6.95

THE AMERICAN SENATOR, Anthony Trollope. Little known, long unavailable Trollope novel on a grand scale. Here are humorous comment on American vs. English culture, and stunning portrayal of a heroine/villainess. Superb evocation of Victorian village life. 561pp. 5⅜ x 8½.
23801-6 Pa. $6.00

WAS IT MURDER? James Hilton. The author of *Lost Horizon* and *Goodbye, Mr. Chips* wrote one detective novel (under a pen-name) which was quickly forgotten and virtually lost, even at the height of Hilton's fame. This edition brings it back—a finely crafted public school puzzle resplendent with Hilton's stylish atmosphere. A thoroughly English thriller by the creator of Shangri-la. 252pp. 5⅜ x 8. (Available in U.S. only)
23774-5 Pa. $3.00

CENTRAL PARK: A PHOTOGRAPHIC GUIDE, Victor Laredo and Henry Hope Reed. 121 superb photographs show dramatic views of Central Park: Bethesda Fountain, Cleopatra's Needle, Sheep Meadow, the Blockhouse, plus people engaged in many park activities: ice skating, bike riding, etc. Captions by former Curator of Central Park, Henry Hope Reed, provide historical view, changes, etc. Also photos of N.Y. landmarks on park's periphery. 96pp. 8½ x 11. 23750-8 Pa. $4.50

NANTUCKET IN THE NINETEENTH CENTURY, Clay Lancaster. 180 rare photographs, stereographs, maps, drawings and floor plans recreate unique American island society. Authentic scenes of shipwreck, lighthouses, streets, homes are arranged in geographic sequence to provide walking-tour guide to old Nantucket existing today. Introduction, captions. 160pp. 8⅞ x 11¾. 23747-8 Pa. $6.95

STONE AND MAN: A PHOTOGRAPHIC EXPLORATION, Andreas Feininger. 106 photographs by *Life* photographer Feininger portray man's deep passion for stone through the ages. Stonehenge-like megaliths, fortified towns, sculpted marble and crumbling tenements show textures, beauties, fascination. 128pp. 9¼ x 10¾. 23756-7 Pa. $5.95

CIRCLES, A MATHEMATICAL VIEW, D. Pedoe. Fundamental aspects of college geometry, non-Euclidean geometry, and other branches of mathematics: representing circle by point. Poincare model, isoperimetric property, etc. Stimulating recreational reading. 66 figures. 96pp. 5⅝ x 8¼.
63698-4 Pa. $2.75

THE DISCOVERY OF NEPTUNE, Morton Grosser. Dramatic scientific history of the investigations leading up to the actual discovery of the eighth planet of our solar system. Lucid, well-researched book by well-known historian of science. 172pp. 5⅜ x 8½. 23726-5 Pa. $3.00

THE DEVIL'S DICTIONARY. Ambrose Bierce. Barbed, bitter, brilliant witticisms in the form of a dictionary. Best, most ferocious satire America has produced. 145pp. 5⅜ x 8½. 20487-1 Pa. $1.75

SECOND PIATIGORSKY CUP, edited by Isaac Kashdan. One of the greatest tournament books ever produced in the English language. All 90 games of the 1966 tournament, annotated by players, most annotated by both players. Features Petrosian, Spassky, Fischer, Larsen, six others. 228pp. 5⅜ x 8½. 23572-6 Pa. $3.50

ENCYCLOPEDIA OF CARD TRICKS, revised and edited by Jean Hugard. How to perform over 600 card tricks, devised by the world's greatest magicians: impromptus, spelling tricks, key cards, using special packs, much, much more. Additional chapter on card technique. 66 illustrations. 402pp. 5⅜ x 8½. (Available in U.S. only) 21252-1 Pa. $3.95

MAGIC: STAGE ILLUSIONS, SPECIAL EFFECTS AND TRICK PHOTOGRAPHY, Albert A. Hopkins, Henry R. Evans. One of the great classics; fullest, most authorative explanation of vanishing lady, levitations, scores of other great stage effects. Also small magic, automata, stunts. 446 illustrations. 556pp. 5⅜ x 8½. 23344-8 Pa. $5.00

THE SECRETS OF HOUDINI, J. C. Cannell. Classic study of Houdini's incredible magic, exposing closely-kept professional secrets and revealing, in general terms, the whole art of stage magic. 67 illustrations. 279pp. 5⅜ x 8½. 22913-0 Pa. $3.00

HOFFMANN'S MODERN MAGIC, Professor Hoffmann. One of the best, and best-known, magicians' manuals of the past century. Hundreds of tricks from card tricks and simple sleight of hand to elaborate illusions involving construction of complicated machinery. 332 illustrations. 563pp. 5⅜ x 8½. 23623-4 Pa. $6.00

MADAME PRUNIER'S FISH COOKERY BOOK, Mme. S. B. Prunier. More than 1000 recipes from world famous Prunier's of Paris and London, specially adapted here for American kitchen. Grilled tournedos with anchovy butter, Lobster a la Bordelaise, Prunier's prized desserts, more. Glossary. 340pp. 5⅜ x 8½. (Available in U.S. only) 22679-4 Pa. $3.00

FRENCH COUNTRY COOKING FOR AMERICANS, Louis Diat. 500 easy-to-make, authentic provincial recipes compiled by former head chef at New York's Fitz-Carlton Hotel: onion soup, lamb stew, potato pie, more. 309pp. 5⅜ x 8½. 23665-X Pa. $3.95

SAUCES, FRENCH AND FAMOUS, Louis Diat. Complete book gives over 200 specific recipes: bechamel, Bordelaise, hollandaise, Cumberland, apricot, etc. Author was one of this century's finest chefs, originator of vichyssoise and many other dishes. Index. 156pp. 5⅜ x 8. 23663-3 Pa. $2.50

TOLL HOUSE TRIED AND TRUE RECIPES, Ruth Graves Wakefield. Authentic recipes from the famous Mass. restaurant: popovers, veal and ham loaf, Toll House baked beans, chocolate cake crumb pudding, much more. Many helpful hints. Nearly 700 recipes. Index. 376pp. 5⅜ x 8½. 23560-2 Pa. $4.00

"OSCAR" OF THE WALDORF'S COOKBOOK, Oscar Tschirky. Famous American chef reveals 3455 recipes that made Waldorf great; cream of French, German, American cooking, in all categories. Full instructions, easy home use. 1896 edition. 907pp. 6⅝ x 9⅜. 20790-0 Clothbd. $15.00

COOKING WITH BEER, Carole Fahy. Beer has as superb an effect on food as wine, and at fraction of cost. Over 250 recipes for appetizers, soups, main dishes, desserts, breads, etc. Index. 144pp. 5⅜ x 8½. (Available in U.S. only) 23661-7 Pa. $2.50

STEWS AND RAGOUTS, Kay Shaw Nelson. This international cookbook offers wide range of 108 recipes perfect for everyday, special occasions, meals-in-themselves, main dishes. Economical, nutritious, easy-to-prepare: goulash, Irish stew, boeuf bourguignon, etc. Index. 134pp. 5⅜ x 8½. 23662-5 Pa. $2.50

DELICIOUS MAIN COURSE DISHES, Marian Tracy. Main courses are the most important part of any meal. These 200 nutritious, economical recipes from around the world make every meal a delight. "I . . . have found it so useful in my own household,"—*N.Y. Times*. Index. 219pp. 5⅜ x 8½. 23664-1 Pa. $3.00

FIVE ACRES AND INDEPENDENCE, Maurice G. Kains. Great back-to-the-land classic explains basics of self-sufficient farming: economics, plants, crops, animals, orchards, soils, land selection, host of other necessary things. Do not confuse with skimpy faddist literature; Kains was one of America's greatest agriculturalists. 95 illustrations. 397pp. 5⅜ x 8½. 20974-1 Pa.$3.95

A PRACTICAL GUIDE FOR THE BEGINNING FARMER, Herbert Jacobs. Basic, extremely useful first book for anyone thinking about moving to the country and starting a farm. Simpler than Kains, with greater emphasis on country living in general. 246pp. 5⅜ x 8½. 23675-7 Pa. $3.50

A GARDEN OF PLEASANT FLOWERS (PARADISI IN SOLE: PARADISUS TERRESTRIS), John Parkinson. Complete, unabridged reprint of first (1629) edition of earliest great English book on gardens and gardening. More than 1000 plants & flowers of Elizabethan, Jacobean garden fully described, most with woodcut illustrations. Botanically very reliable, a "speaking garden" of exceeding charm. 812 illustrations. 628pp. 8½ x 12¼. 23392-8 Clothbd. $25.00

ACKERMANN'S COSTUME PLATES, Rudolph Ackermann. Selection of 96 plates from the *Repository of Arts*, best published source of costume for English fashion during the early 19th century. 12 plates also in color. Captions, glossary and introduction by editor Stella Blum. Total of 120pp. 8⅜ x 11¼. 23690-0 Pa. $4.50

MUSHROOMS, EDIBLE AND OTHERWISE, Miron E. Hard. Profusely illustrated, very useful guide to over 500 species of mushrooms growing in the Midwest and East. Nomenclature updated to 1976. 505 illustrations. 628pp. 6½ x 9¼. 23309-X Pa. $7.95

AN ILLUSTRATED FLORA OF THE NORTHERN UNITED STATES AND CANADA, Nathaniel L. Britton, Addison Brown. Encyclopedic work covers 4666 species, ferns on up. Everything. Full botanical information, illustration for each. This earlier edition is preferred by many to more recent revisions. 1913 edition. Over 4000 illustrations, total of 2087pp. 6⅛ x 9¼. 22642-5, 22643-3, 22644-1 Pa., Three-vol. set $24.00

MANUAL OF THE GRASSES OF THE UNITED STATES, A. S. Hitchcock, U.S. Dept. of Agriculture. The basic study of American grasses, both indigenous and escapes, cultivated and wild. Over 1400 species. Full descriptions, information. Over 1100 maps, illustrations. Total of 1051pp. 5⅜ x 8½. 22717-0, 22718-9 Pa., Two-vol. set $12.00

THE CACTACEAE,, Nathaniel L. Britton, John N. Rose. Exhaustive, definitive. Every cactus in the world. Full botanical descriptions. Thorough statement of nomenclatures, habitat, detailed finding keys. The one book needed by every cactus enthusiast. Over 1275 illustrations. Total of 1080pp. 8 x 10¼. 21191-6, 21192-4 Clothbd., Two-vol. set $35.00

AMERICAN MEDICINAL PLANTS, Charles F. Millspaugh. Full descriptions, 180 plants covered: history; physical description; methods of preparation with all chemical constituents extracted; all claimed curative or adverse effects. 180 full-page plates. Classification table. 804pp. 6½ x 9¼. 23034-1 Pa. $10.00

A MODERN HERBAL, Margaret Grieve. Much the fullest, most exact, most useful compilation of herbal material. Gigantic alphabetical encyclopedia, from aconite to zedoary, gives botanical information, medical properties, folklore, economic uses, and much else. Indispensable to serious reader. 161 illustrations. 888pp. 6½ x 9¼. (Available in U.S. only) 22798-7, 22799-5 Pa., Two-vol. set $11.00

THE HERBAL or GENERAL HISTORY OF PLANTS, John Gerard. The 1633 edition revised and enlarged by Thomas Johnson. Containing almost 2850 plant descriptions and 2705 superb illustrations, Gerard's *Herbal* is a monumental work, the book all modern English herbals are derived from, the one herbal every serious enthusiast should have in its entirety. Original editions are worth perhaps $750. 1678pp. 8½ x 12¼. 23147-X Clothbd. $50.00

MANUAL OF THE TREES OF NORTH AMERICA, Charles S. Sargent. The basic survey of every native tree and tree-like shrub, 717 species in all. Extremely full descriptions, information on habitat, growth, locales, economics, etc. Necessary to every serious tree lover. Over 100 finding keys. 783 illustrations. Total of 986pp. 5⅜ x 8½. 20277-1, 20278-X Pa., Two-vol. set $10.00

AMERICAN BIRD ENGRAVINGS, Alexander Wilson et al. All 76 plates. from Wilson's *American Ornithology* (1808-14), most important ornithological work before Audubon, plus 27 plates from the supplement (1825-33) by Charles Bonaparte. Over 250 birds portrayed. 8 plates also reproduced in full color. 111pp. 9⅜ x 12½. 23195-X Pa. $6.00

CRUICKSHANK'S PHOTOGRAPHS OF BIRDS OF AMERICA, Allan D. Cruickshank. Great ornithologist, photographer presents 177 closeups, groupings, panoramas, flightings, etc., of about 150 different birds. Expanded *Wings in the Wilderness*. Introduction by Helen G. Cruickshank. 191pp. 8¼ x 11. 23497-5 Pa. $6.00

AMERICAN WILDLIFE AND PLANTS, A. C. Martin, et al. Describes food habits of more than 1000 species of mammals, birds, fish. Special treatment of important food plants. Over 300 illustrations. 500pp. 5⅜ x 8½. 20793-5 Pa. $4.95

THE PEOPLE CALLED SHAKERS, Edward D. Andrews. Lifetime of research, definitive study of Shakers: origins, beliefs, practices, dances, social organization, furniture and crafts, impact on 19th-century USA, present heritage. Indispensable to student of American history, collector. 33 illustrations. 351pp. 5⅜ x 8½. 21081-2 Pa. $4.00

OLD NEW YORK IN EARLY PHOTOGRAPHS, Mary Black. New York City as it was in 1853-1901, through 196 wonderful photographs from N.-Y. Historical Society. Great Blizzard, Lincoln's funeral procession, great buildings. 228pp. 9 x 12. 22907-6 Pa. $7.95

MR. LINCOLN'S CAMERA MAN: MATHEW BRADY, Roy Meredith. Over 300 Brady photos reproduced directly from original negatives, photos. Jackson, Webster, Grant, Lee, Carnegie, Barnum; Lincoln; Battle Smoke, Death of Rebel Sniper, Atlanta Just After Capture. Lively commentary. 368pp. 8⅜ x 11¼. 23021-X Pa. $8.95

TRAVELS OF WILLIAM BARTRAM, William Bartram. From 1773-8, Bartram explored Northern Florida, Georgia, Carolinas, and reported on wild life, plants, Indians, early settlers. Basic account for period, entertaining reading. Edited by Mark Van Doren. 13 illustrations. 141pp. 5⅜ x 8½. 20013-2 Pa. $4.50

THE GENTLEMAN AND CABINET MAKER'S DIRECTOR, Thomas Chippendale. Full reprint, 1762 style book, most influential of all time; chairs, tables, sofas, mirrors, cabinets, etc. 200 plates, plus 24 photographs of surviving pieces. 249pp. 9⅞ x 12¾. 21601-2 Pa. $6.50

AMERICAN CARRIAGES, SLEIGHS, SULKIES AND CARTS, edited by Don H. Berkebile. 168 Victorian illustrations from catalogues, trade journals, fully captioned. Useful for artists. Author is Assoc. Curator, Div. of Transportation of Smithsonian Institution. 168pp. 8½ x 9½. 23328-6 Pa. $5.00

THE SENSE OF BEAUTY, George Santayana. Masterfully written discussion of nature of beauty, materials of beauty, form, expression; art, literature, social sciences all involved. 168pp. 5⅜ x 8½. 20238-0 Pa. $2.50

ON THE IMPROVEMENT OF THE UNDERSTANDING, Benedict Spinoza. Also contains *Ethics, Correspondence,* all in excellent R. Elwes translation. Basic works on entry to philosophy, pantheism, exchange of ideas with great contemporaries. 402pp. 5⅜ x 8½. 20250-X Pa. $4.50

THE TRAGIC SENSE OF LIFE, Miguel de Unamuno. Acknowledged masterpiece of existential literature, one of most important books of 20th century. Introduction by Madariaga. 367pp. 5⅜ x 8½. 20257-7 Pa. $3.50

THE GUIDE FOR THE PERPLEXED, Moses Maimonides. Great classic of medieval Judaism attempts to reconcile revealed religion (Pentateuch, commentaries) with Aristotelian philosophy. Important historically, still relevant in problems. Unabridged Friedlander translation. Total of 473pp. 5⅜ x 8½. 20351-4 Pa. $5.00

THE I CHING (THE BOOK OF CHANGES), translated by James Legge. Complete translation of basic text plus appendices by Confucius, and Chinese commentary of most penetrating divination manual ever prepared. Indispensable to study of early Oriental civilizations, to modern inquiring reader. 448pp. 5⅜ x 8½. 21062-6 Pa. $4.00

THE EGYPTIAN BOOK OF THE DEAD, E. A. Wallis Budge. Complete reproduction of Ani's papyrus, finest ever found. Full hieroglyphic text, interlinear transliteration, word for word translation, smooth translation. Basic work, for Egyptology, for modern study of psychic matters. Total of 533pp. 6½ x 9¼. (Available in U.S. only) 21866-X Pa. $4.95

THE GODS OF THE EGYPTIANS, E. A. Wallis Budge. Never excelled for richness, fullness: all gods, goddesses, demons, mythical figures of Ancient Egypt; their legends, rites, incarnations, variations, powers, etc. Many hieroglyphic texts cited. Over 225 illustrations, plus 6 color plates. Total of 988pp. 6⅛ x 9¼. (Available in U.S. only)
22055-9, 22056-7 Pa., Two-vol. set $12.00

THE ENGLISH AND SCOTTISH POPULAR BALLADS, Francis J. Child. Monumental, still unsuperseded; all known variants of Child ballads, commentary on origins, literary references, Continental parallels, other features. Added: papers by G. L. Kittredge, W. M. Hart. Total of 2761pp. 6½ x 9¼.
21409-5, 21410-9, 21411-7, 21412-5, 21413-3 Pa., Five-vol. set $37.50

CORAL GARDENS AND THEIR MAGIC, Bronsilaw Malinowski. Classic study of the methods of tilling the soil and of agricultural rites in the Trobriand Islands of Melanesia. Author is one of the most important figures in the field of modern social anthropology. 143 illustrations. Indexes. Total of 911pp. of text. 5⅝ x 8¼. (Available in U.S. only)
23597-1 Pa. $12.95

AN AUTOBIOGRAPHY, Margaret Sanger. Exciting personal account of hard-fought battle for woman's right to birth control, against prejudice, church, law. Foremost feminist document. 504pp. 5⅜ x 8½.
20470-7 Pa. $5.50

MY BONDAGE AND MY FREEDOM, Frederick Douglass. Born as a slave, Douglass became outspoken force in antislavery movement. The best of Douglass's autobiographies. Graphic description of slave life. Introduction by P. Foner. 464pp. 5⅜ x 8½. 22457-0 Pa. $5.00

LIVING MY LIFE, Emma Goldman. Candid, no holds barred account by foremost American anarchist: her own life, anarchist movement, famous contemporaries, ideas and their impact. Struggles and confrontations in America, plus deportation to U.S.S.R. Shocking inside account of persecution of anarchists under Lenin. 13 plates. Total of 944pp. 5⅜ x 8½.
22543-7, 22544-5 Pa., Two-vol. set $9.00

LETTERS AND NOTES ON THE MANNERS, CUSTOMS AND CONDITIONS OF THE NORTH AMERICAN INDIANS, George Catlin. Classic account of life among Plains Indians: ceremonies, hunt, warfare, etc. Dover edition reproduces for first time all original paintings. 312 plates. 572pp. of text. 6⅛ x 9¼. 22118-0, 22119-9 Pa.. Two-vol. set $10.00

THE MAYA AND THEIR NEIGHBORS, edited by Clarence L. Hay, others. Synoptic view of Maya civilization in broadest sense, together with Northern, Southern neighbors. Integrates much background, valuable detail not elsewhere. Prepared by greatest scholars: Kroeber, Morley, Thompson, Spinden, Vaillant, many others. Sometimes called Tozzer Memorial Volume. 60 illustrations, linguistic map. 634pp. 5⅜ x 8½.
23510-6 Pa. $7.50

HANDBOOK OF THE INDIANS OF CALIFORNIA, A. L. Kroeber. Foremost American anthropologist offers complete ethnographic study of each group. Monumental classic. 459 illustrations, maps. 995pp. 5⅜ x 8½.
23368-5 Pa. $10.00

SHAKTI AND SHAKTA, Arthur Avalon. First book to give clear, cohesive analysis of Shakta doctrine, Shakta ritual and Kundalini Shakti (yoga). Important work by one of world's foremost students of Shaktic and Tantric thought. 732pp. 5⅜ x 8½. (Available in U.S. only)
23645-5 Pa. $7.95

AN INTRODUCTION TO THE STUDY OF THE MAYA HIEROGLYPHS, Syvanus Griswold Morley. Classic study by one of the truly great figures in hieroglyph research. Still the best introduction for the student for reading Maya hieroglyphs. New introduction by J. Eric S. Thompson. 117 illustrations. 284pp. 5⅜ x 8½. 23108-9 Pa. $4.00

A STUDY OF MAYA ART, Herbert J. Spinden. Landmark classic interprets Maya symbolism, estimates styles, covers ceramics, architecture, murals, stone carvings as artforms. Still a basic book in area. New introduction by J. Eric Thompson. Over 750 illustrations. 341pp. 8⅜ x 11¼.
21235-1 Pa. $6.95